Robotics : An Introduction

Open University Press Robotics Series

Edited by

P. G. Davey   CBE MA MIEE MBCS C.Eng

This series is designed to give undergraduate, graduate and practising engineers access to this fast developing field and provide an understanding of the essentials both of robot design and of the implementation of complete robot systems for CIM and FMS. Individual titles are oriented either towards industrial practice and current experience or towards those areas where research is actively advancing to bring new robot systems and capabilities into production.

The design and overall editorship of the series are due to Peter Davey, Managing Director of Meta Machines Limited, Abingdon; Fellow of St Cross College, Oxford University; and formerly Co-ordinator of the UK Science and Engineering Research Council's Programme in Industrial Robotics.

His wide ranging responsibilities and international involvement in robotics research and development endow the series with unusual quality and authority.

TITLES IN THE SERIES

Robotics: An Introduction	D. McCloy and M. Harris
Robots in Assembly	A. Redford and E. Lo

Titles in preparation

Industrial Robotics Applications	E. Appleton and D. Williams
Integration of Robots within Manufacturing Systems	R. Weston, C. Sumpter and J. Gascoigne
Unsolved Problems in Robotics	R. Popplestone
Sensors and Processing for Robots	R. Ruocco

Robotics : An Introduction

D. McCloy
D. M. J. Harris

OPEN UNIVERSITY PRESS

Milton Keynes

Open University Press
Open University Educational Enterprises Limited
12 Cofferidge Close
Stony Stratford
Milton Keynes MK11 1BY, England

First Published 1986

British Library Cataloguing in Publication Data

McCloy, D.
   Robotics: an introduction. — (Robotics series)
   1. Robots
   I. Title   II. Harris, D.M.J.   III.
   Series
   629.8′92   TJ211

ISBN 0-335-15402-6

ISBN 0-335-15401-8 Pbk

Text design by Clarke Williams

Typeset and printed in Great Britain by The Universities Press (Belfast) Ltd.

Contents

Series Editor's Preface

All too often, it seems to me, books on robotics tend to strand the reader just at the point where 'hand waving' accounts stop and real engineering design begins. That is, the engineer reaches an understanding of the need for each element in a robot system but then cannot find a proper review of the devices and techniques which are actually used to provide those features according to today's best engineering practice.

In this book, Professor McCloy and Mr Harris have worked from the most general concepts of robotics, through a clear exposition of the essential mathematics of moving linkages and their control, to an explanation of the key features of actuation, perception, 'intelligence' and manufacturing applications for robots. It is their considerable achievement to have provided in a single book summary accounts of engineering practice in fields as diverse as, for instance, electric actuators, proprioceptive and exteroceptive sensors, basic control theory, walking devices, teleoperators and robotic vision. This saves the student or industrial engineer from having to dig at once into the special literature particular to each field—an excavation which is not only difficult and time consuming but which also obscures the attractive unity of robotics studies.

The authors' achievement makes it particularly fitting that the book is one of the first two titles to be published in our Open University series. In fact it serves as an introduction to the architecture of the entire series, in that other titles we have commissioned will describe in greater depth how each of these fields relates to robotics and robot applications.

I hope this book will be widely used, as an overview and as a text book, and as a remarkably comprehensive work of reference for engineers working in the field of robotics for many years to come.

P. G. Davey

Introduction

Industrial robots have undergone many developments since the arrival of the first Unimation machine in 1962. Applications are widespread and growing. The robot's advantages of flexibility, reprogrammability, tirelessness and hardiness have come to be appreciated by industrialists, and even the layman realizes that today's industrial robot, unlike the tin marvel of science fiction, has a real and useful role to fulfil.

However, in spite of many developments in the associated technologies, the industrial robot has yet to achieve its full potential. A second generation of robots capable of sensing and reacting to the external environment is still in its infancy. Robots have been and continue to be shaped in the image of man, and the second generation is seen to be a further step in this direction. But anthropomorphism is a constraining influence, and new perspectives are needed. Perhaps we should endeavour to see beyond the principle of replacing a human by a human-like robot. The time is ripe for a new look at manufacturing processes, particularly assembly, and including product design for assembly. Flexible manufacturing systems involving squat multi-armed robots with second-generation capabilities, interacting with machine tools, and all under integrated computer control, offer great potential. Furthermore, as robot capabilities are enhanced to meet the challenge of the more difficult industrial tasks, applications outside the factory will become more feasible, lending credence to the present research on mobile robots capable of operating in unstructured environments such as homes and hospitals.

Robotics is a multidisciplinary activity: it is an ideal vehicle for illustrating the systems approach and as such it offers a useful means of broadening a student's perspectives. Because of this we think it unwise to adopt an over-rigid approach to the subject, bound by conventional definitions. The closely related technologies of pick-and-place devices, walking machines, tele-operators and prosthetics cannot be ignored.

The engineer and manager in industry, and the engineering student, require comprehensive yet detailed information on the design and application of all elements of automated manufacturing, and this book is intended to fulfil this need in the area of industrial robots. As with most computer-controlled equipment, it is possible to operate an industrial robot successfully with very little knowledge of the technologies involved in its sub-systems. In order to do so the operator requires familiarity with the teaching and communication procedures applicable to the particular machine, information which can be gleaned from instruction manuals and from training courses. But to select the best robot for a particular task, and to apply it to that task, requires more than this. It requires an insight into the capabilities of the many types of machines available, which in turn requires a sound understanding of the many technologies contributing to an industrial robot system.

The content of this book is a logical outcome of the influences mentioned above. Chapter 1 sets the scene by defining the topics of interest and exploring their relevance in the wider field of automation. Chapter 2 shows how mechanism theory relates to robots, and deals with manipulator configurations, particularly with regard to the mathematics of linkages. Wrists and end-effectors are considered in Chapter 3, and also the technology of walking machines. The fundamental areas of actuation, control, measurement and computers are covered in Chapters 4, 5, 6 and 7. Chapter 8 discusses the interaction of the robot with its environment; an examination of robot vision leads to pattern recognition and the role of artificial intelligence. Some of the more common industrial applications of robots are detailed in Chapter 9. The many uses of tele-operators and associated devices are considered in Chapter 10. Finally, Chapters 11 and 12 look at the financial aspects of robots, and consider some of the social consequences of the technology.

Because this book builds on such a variety of technologies, it is felt that greatest advantage will be derived from it in relation to final-year undergraduate and postgraduate programmes. Industrialists may also benefit, not only through a closer knowledge of the technology but, just as importantly, through a feel for applications, costs and benefits.

The authors have drawn heavily on their experiences of teaching the subject to students and industrialists at the University of Ulster. The book was not easy to write, but the authors were fortunate to have the expert assistance of Estelle Goyer, who did the typing (and advised on spelling), and Reg McClure whose artistic efforts added a lot to the presentation.

Don McCloy
Michael Harris

List of abbreviations and acronyms

A/D	analogue-to-digital (conversion)	CP	continuous-path (control)
AC	alternating current	CPU	central processing unit
ADC	analogue-to-digital converter	CRT	cathode-ray tube
AGV	automatic guided vehicle	D/A	digital-to-analogue (conversion)
ASCII	American Standard Code for Information Interchange	DAC	digital-to-analogue converter
ASR	automatic speech recognition	DC	direct current
		DIL	dual-in-line (switches)
AVR	automatic voltage regulator	DNC	direct numerical control
BASIC	a high-level computer language: Beginners' All-purpose Symbolic Code	EAROM	electrically alterable read-only memory
		ECL	emitter coupled logic
		EPROM	erasable programmable read-only memory
CAD	computer-aided design	FFTO	free-flying tele-operator
CAM	computer-aided manufacture	FMS	flexible manufacturing system
CID	control input device	FORTRAN	A high-level computer language
CMOS	complementary MOS	I/O	input/output
CNC	computerized numerical control	IIL	integrated injection logic

IWR	isolated word recognition system	RAM	random access memory
LCD	liquid crystal display	RCC	remote centre compliance
LED	light-emitting diode		
LIDAR	light detection and ranging	RMRC	resolved motion rate control
LSB	least significant bit	ROM	read-only memory
LSI	large-scale integration	RVDT	rotary variable differential transformer
LVDT	linear variable differential transformer		
		SATO	shuttle-attached teleoperator
MIG	inert gas-metal-arc	TCP	tool centre point
MOS	metal oxide silicon	TIG	inert gas- tungsten-arc
MPU	microprocessor unit		
MSB	most significant bit	TTL	transistor-transistor logic
NC	numerical control		
PID	proportional integral derivative	VAL	a high-level language developed by Unimation for their robots
PLC	programmable logic controller		
PROM	programmable read-only memory	VDU	visual display unit
		VLSI	very large scale integration
PTP	point-to-point (control)		

Chapter 1

From flint tool to flexible manufacture

1.1 Introduction

Technology is as old as Man. People first became technologists when they learned to take advantage of the materials and natural phenomena of the physical world. When they discovered that a bone or stick could be used to kill animals and to move rocks they became toolmakers, and tools are the trademark of the technologist. Tools, from sticks to tele-operators, have made the human race pre-eminent in the animal kingdom. Thomas Carlyle described this nicely: 'Man stands on a basis, at most of the flattest soled, of some half square foot insecurely enough. Three quintals are a crushing load for him; the steer of the field tosses him aloft like a waste rag. Nevertheless he can use tools. Without tools he is nothing. With tools he is all'. Tools have increased human capability but they also indicate another important difference between us and the animals—the concept of planning. When one makes a tool one has a use in mind: planning and problem-solving are central to technology.

1.2 Technology extends human capabilities

This process of problem-solving and applying knowledge has led, in an evolutionary manner, to a variety of technologies, each extending human capabilities in a particular way (McCloy, 1984) (Figure 1.1). Muscle power has been extended by machines and mechanization: brain power has been augmented by the computer: our senses have been extended by instruments and measuring devices: our capabilities of control have been enhanced by

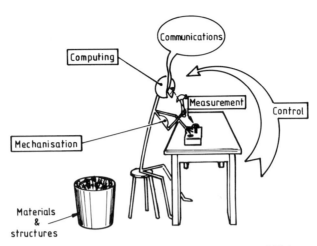

Figure 1.1 Technology extends man's capabilities.

cybernetics: our speed and range of communication have been increased enormously by telecommunications: and the store of materials and structures available for artifacts is bountiful, including ceramics, carbon fibres and composites.

We have now reached an exciting and dramatic era when technology, particularly robot technology, will not only extend our human capabilities but may well replace them entirely. But before pursuing that topic further it would be useful to take a closer look at these capabilities.

The quotation from Carlyle emphasized our physical puniness in relation to the other animals. One of the world's strongest men, Paul Anderson, an Olympic weightlifting champion, was able to lift a weight of about 25 kN, equivalent to 38 people. But this is tiny compared to many of today's industrial tasks. And what about power? Experiments have shown that for intervals less than 1 s a very fit man can generate about 1 kW, but as the length of the working period goes up the power output reduces to 200 W over 10 s and about 100 W for longer periods. This is small in comparison to the average car's 50 kW.

Our senses have their limitations too. Hearing is restricted to the frequency range 20–20 000 Hz, and pressure amplitudes in excess of 200 N/m^2 (one five-hundredth of an atmosphere) can cause deafness. Vision, too, is limited. The spectrum of electromagnetic radiation extends from wavelengths of around 0.001 nm (gamma rays) to around 100 km (long radio waves), but within this broad range the eye is restricted to the narrow band from 390 nm (violet light) to 750 nm (red light).

Another 'shortcoming' of our senses is the need for a sizeable change in a stimulus before our sense organs can recognize that a change has in fact occurred. This phenomenon was thoroughly researched by E. H. Weber in 1834; he defined the Weber fraction as the ratio of the change in stimulus required before the human perceives that it has in fact changed, to the original stimulus.

Table 1.1

Sense	Weber fraction
Vision (brightness, white light)	1/60
Kinesthesis (lifted weights)	1/50
Pain (thermal on skin)	1/30
Hearing (middle pitch, moderate loudness)	1/10
Pressure	1/7
Smell (odour of India rubber)	1/4
Taste (table salt)	1/3

For example, a mass of 10 kg would have to be increased by about 0.2 kg before the person holding it would notice any change. The table shows that we are best at discriminating brightness changes and worst at discriminating taste changes. Thus whilst our sensory apparatus has evolved over millions of years to a degree of perfection suited to our normal environment, it cannot meet the demands of today's world where science, technology and commerce require many aspects of the physical world to be determined extremely accurately.

Our communication ability is also limited both in range and in content. Unaided, we can transmit about one word per second over a distance of about 2 km. Nowadays telecommunications have increased the range of human communication to effectively limitless extremes, but there is still the problem of rate of exchange of information. Information is measured in terms of the least and simplest unit, which is the transmission of a single decision between equally probable alternatives, such as heads or tails, yes or no, or on or off. This is ideally suited to the binary system, 1 perhaps representing yes and 0 representing no. Thus one bit (binary digit) is the basic unit of information. For example, if you are asked the sex of your only child and you reply female, then you have given one bit of information. Again, if there are 64 houses in your street and you tell someone which one you live in, then you have given six bits of information, for in order to extract that information it would have been necessary to have asked at least six questions with yes/no answers.

The human brain can comfortably accept and cope with around 25 bits per second. It can be shown that a word of the English language has an information content of around five bits, and the average novel about 250 000 bits. So you should be able to read the average novel in about 10 000 s, at 25 bits/s. But compare this with the rate at which information is displayed on a black and white television screen—37.5×10^6 bits/s.

All of these limitations, combined with our physical frailty and inability to work in hostile environments, have encouraged the growth of mechanization and automation.

1.3 Mechanization

Since the forces generated by humans are relatively low it became necessary
to invent amplifying devices in the form of machines; the five most
elementary ones, enumerated by Hero of Alexandria, are the lever, the
wheel and axle, the pulley, the wedge and the screw. The word machine is
notoriously difficult to define, and the Webster definition is as good as any:
'a machine is the intermediary between the motive power and those parts
which actually carry out the movements necessary to the required work'.
Indeed, when a human provides the motive power the intermediary function
of the machine is to transform the human motive power into a form that
matches the requirements of the workpiece. People can generate sufficient
energy for many tasks, but often the energy is not in the right form. For
example, you would have no difficulty in climbing a 50 cm step, and if you
weighed 600 N you would have used up 300 J of energy. Now although the
same amount of energy would raise a weight of 6 kN through 5 cm, the
magnitude of the force involved would place this task beyond human
capability. A machine is needed to match our energy to this task. A lever
with a 20:1 ratio could reduce the human effort to a manageable 300 N, the
penalty being that, in order to move the load through the required 5 cm, the
operator would have to move the control end of the lever through 100 cm.
But this is an acceptable penalty, since such a task is within human
capability. Thus the machine acts as an energy transformer, transforming
energy into a form suitable to a given task.

The transformation abilities of machines are not infinite, however, and
there are many occasions when the limitations on human capability present
the machine with too great a challenge. In such cases human muscle power
has to be replaced by animals or engines. Engines, or prime movers, convert
a naturally occurring form of energy into a more directly useful form of
work. The windmill, one of the earliest prime movers, converted the
naturally occurring kinetic energy of moving air into the kinetic energy of a
turning grindstone. The internal combustion engine converts the naturally
occurring chemical energy of oil into the kinetic energy of a rotating
flywheel. The invention of such engines led us into the age of mechanization.

Mechanization, or the use of machines to do the work of animals or
people, has been with us for centuries. Its growth was particularly rapid
during the Industrial Revolution, when the introduction of steam power
allowed many manual operations to be relegated to the machine. But,
although reducing the physical effort, mechanization still left us with the
burden of control. People had to feed, guide and correct the movements of
their machines. The next evolutionary step in technological development,
automation, was to take this burden of control from us and give it to the
machine.

1.4 Automatic control

Before attempting to define automation we should clarify our thoughts on control, particularly automatic control. Automatic control is self-acting control—control without human intervention. The large majority of today's industrial control systems could not function if they relied on human operators. Our inbuilt reaction time of about 0.2 s, our tendency to become bored and to be distracted, are among the factors that exclude us from such systems.

Automatic control falls into two major subsections: control of the order of events and control of physical variables. Let us look at these in turn.

Controlling the order of events

Many manufacturing processes require the order of events to be controlled. For example, a drilling operation may include the following sequence of events: (1) the component is pushed into position, (2) the component is clamped, (3) the component is drilled, (4) the clamp is released and (5) the component is removed. Such a task is easy for the human operator but it is exceedingly repetitive; the operator becomes bored and performance can degrade. However, since there is a clearly defined programme of events it is relatively easy to convert this to an automatic or self-acting system. The wonderful automata of the nineteenth century and the robots of today have a feature in common with this system—they all have to be capable of being programmed to follow a desired sequence of events. An automaton may have to lift a pen, turn its head, roll its eyes and set the pen down again. A robot may have to lift a hot component, turn it over and place it in a stamping machine. How do we tell these machines what we want them to do? How do we *program* them?

In the earliest automata, clockwork mechanisms ensured the proper timing of each event. The principle of operation can be seen in the musical box which uses a rotary drum whose surface is covered with tiny pins. As the drum turns the pins strike at tiny comb-like teeth, each with its own pitch, and the sequence or ordered arrangement of the pins produces the desired tune. The drum and pins constitute the program. Of course in automata, robots and other forms of automatic machines, it is necessary to produce an ordered sequence of *events*, not just of notes; but the principle is the same. In many cases a camshaft is used to control the movements of an automaton. Instead of pins on a rotating drum, the camshaft program uses cams on a rotating shaft (Figure 1.2(a)). The raised portions on the cams (the lobes) make contact with switches or valves which cause actuators to extend and retract, and the program, or the timing of events, can be varied by altering the relative angular positions of the cams.

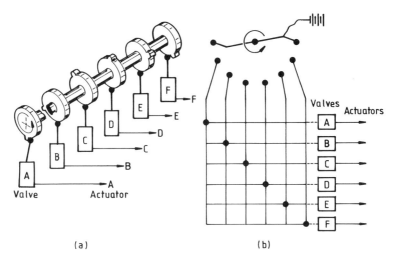

Figure 1.2 Controlling the order of events: (a) the camshaft programmer (b) uniselector and plugboard.

The ease of programming, the *programmability*, is an important feature of automatic sequence control. If programs have to be changed regularly, then the camshaft programmer is not to be recommended: its programmability is low. Adjustment of cams is tedious and time-consuming, so on the whole this form of control is restricted to systems with fixed sequences, such as the operation of engine valve gear.

The use of uniselectors and plugboards can improve programmability (Figure 1.2(b)). As shown, the uniselector steps at a constant rate from contact to contact—six in total. When a contact is made the relevant column of the plugboard becomes live. The plugs connect the columns and rows of the plugboard so, for the given plug configuration, valve *A* will be energized first, followed by *B* and so on. The sequence is changed by rearranging the plugs.

Better programmability is obtained from systems that employ punched cards or tapes to store the instructions. Movement of such cards past a reading head can initiate the appropriate sequence of actions—an old idea, the brainchild of Vaucanson, renowned for his automata. Holes or slots can allow light through to operate photoelectric devices, or needles to trip mechanically operated switches or feelers to make electrical contacts. Indeed, in the earlier 'roller' pianos jets of air passed through the holes to trigger the movement of the hammers. The programmability of such systems is good, since it is a relatively easy matter to replace one card or tape by another, or to punch a new card or tape.

The systems just described, whether camshafts, plugboards or moving tapes, are *time-based controllers*. The camshaft rotates at constant speed, like the hands of a clock, and the timing of each event is determined by the relative angular position of the cams. The uniselector arm steps at a constant rate. In the case of punched tape, the tape can be drawn past a reading head at constant speed or in regularly spaced increments of time. In all three cases the controller moves on inexorably, calling into play the various actions demanded by the program. But how does such a system know that the actions called for do in fact take place? Consider for example the sequence described earlier of feeding, clamping and drilling a component. The sequence could be driven by a camshaft, in which case the various events would be initiated at predetermined times—say clamp 10 s after the feed, and drill 10 s after clamp. But what happens if a component jams whilst being fed in? It would not arrive in the clamp in time. The camshaft would turn relentlessly, and 10 s after the clamp had operated the drill would descend and perhaps put a hole in the clamp or in the base of the machine. Such occurrences can be avoided by using event-based rather than time-based controllers.

Event-based controllers use the completion of one event as the trigger to start the next event. The principle of operation is illustrated in Figure 1.3. An operator pushes button *a*; this energizes actuator *A*, which extends to press button *b*, which operates actuator *B*, and so on. A chain reaction is started; completion of the chain depends on the completion of each individual link. If such an event-based controller were used in our feed–clamp–drill application, then the calamity described earlier could not have occurred. The jammed component would have stopped the feed actuator before the completion of its task—that event would not have been complete and thus the next event, that of clamping, would not have been initiated. The sequence would stop at the point where the problem arose and the operator would then be called upon to correct the error.

Event-based controllers are superior to time-based controllers because they have some idea of how well they are carrying out the particular task. An event-based program comes to a halt when its sensors (the pushbuttons of Figure 1.3) fail to inform it of successful completion of any part of the sequence. Information is being fed back from the task to the controller, and this is the first step towards the intelligent machine.

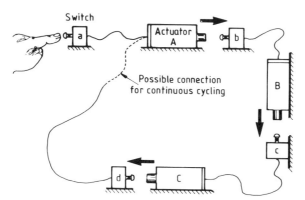

Figure 1.3 Event-based control.

Controlling physical variables

When a controller has no knowledge of the results of its own actions, it is known as an *open-loop controller*. When information about its actions is fed back from the task the controller becomes a *closed-loop controller*. Thus, as far as sequence controls are concerned, the time-based controller can be classified as open-loop, and the event-based controller as closed-loop. But, on the whole, this terminology is more often associated with the control of physical variables such as speed, light intensity or temperature, rather than the control of the order of events.

The concept of closed-loop control is further clarified by reference to an example. Consider the control of the speed of a steam turbine. When such a turbine is being installed the engineer could measure turbine speed and steam valve setting and draw a graph showing their relationship. For subsequent use the technicians in charge could refer to this to determine the valve setting for a particular desired speed. But what would happen if the steam pressure were low, if the load on the turbine increased, or if dirt particles restricted the control valve passages on that particular day? Clearly the turbine speed would differ from the desired value. The control system would have failed in its task: it would have failed because it has no knowledge of the results of its actions. Such systems are called open-loop, because there is no loop of information back from the output.

The loop can be closed by measuring the results of the controller's actions and using those measurements to influence its further actions. In the chosen example this could be done by installing a speed-measuring device on the output shaft, and by employing a human operator to read this meter and to make appropriate adjustments to the control valve. A further development leading to automatic control could be the replacement of the human by a mechanical or electrical device. James Watt, one of the forerunners of automatic control, devised the flyball governor to do such a task in the control system of the early steam engines. Rotating balls were geared to the output shaft. If the speed increased they moved outwards and upwards under the action of centrifugal forces and this movement lifted a sleeve which closed the steam valve. Reduction in speed had the opposite effect.

The block diagram of Figure 1.4 shows the closed loop and its most

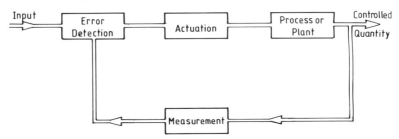

Figure 1.4   Block diagram of a closed-loop system.

important constituent elements—measurement, error detection and actuation. In the speed control example the input was the desired speed and the controlled quantity was the output or actual speed. The process or plant was the turbine; load, and measurement, error detection and actuation involved either a human operator or, in the automatic case, a device such as a flyball governor. Error detection requires the output to be subtracted from the input, which has led to this form of feedback being known as *negative feedback*.

There are three major classifications of negative feedback systems: *servomechanisms, process controls* and *regulators*. Servomechanisms are used to make the motion of an output member follow that of an input member with power amplification. An example is the control of an automatic lathe where the output is the position of the cutting tool and the input could be derived from a template cut to the desired shape, or perhaps from a computer tape. Whilst servomechanisms are restricted to controlling the kinematic variables, process controls cover a much wider field including control of pressure, acidity, temperature, etc. Finally, the class of regulators is restricted to systems with fixed inputs, such as voltage level control.

The main advantage of closed-loop control is the reduction in the effect of disturbing influences, such as increased load, reduced steam pressure and so on. A major disadvantage is the possibility of instability, a situation in which the controlled variable tends to run away, and the loop is, in a sense, chasing its own tail.

Complementary human and machine qualities

So far a lot of attention has been given to human shortcomings and how they have excluded human intervention from the control loop. Let us not forget, however, that people are creative, they can improvise, set priorities and establish values: they can reason. Weighing these advantages against the disadvantages stated earlier, it is possible to assign control orientated tasks to humans and machines. This has been done by Johnsen and Corliss (1969) as shown in Table 1.2.

Table 1.2

Tasks suited to people	Tasks suited to the machine
Pattern recognition	Monitoring multichannel input
Target identification	Boring, repetitive manipulation
New, exploratory manipulation	Precision motion and force applications
Long-term memory	High-speed motions
Trouble-shooting, emergency operation	Short-term memory
Hypothesizing, ideation, planning	Computing
Interpreting variable-format data	Monitoring
Inductive thinking	Deductive analysis
Setting goals, priorities, evaluating results	Development of optimal strategies
	Non-anthropomorphic motions

1.5 Automation

It was necessary to clarify our views on automatic control before moving on to automation, which is an even more complicated concept. The word automation is an elision of 'automatic motivation', and was first used in the 1940s by a Ford Motor Company engineer to describe the collective operation of many interconnected machines in their Detroit plant. The machines were able to mill, drill, hone and tap a rough engine-block casting, ejecting the finished product at the end of the line. As well as carrying out the machining operations, the system was programmed to perform the unproductive tasks of clamping and handling which were formerly done manually. Human effort was required only for supervising the machines and checking the quality of the product.

A clear and unambiguous definition of automation has been elusive. The authors prefer that given in the *Encyclopaedia Britannica*: automation is defined there as 'the performance of automatic operations directed by programmed commands with automatic measurement of action, feedback and decision-making.'

This definition indicates that automation involves a program for determining the order of events as well as instructing the system how each step of the operation is to be carried out. The computer offers the most flexible form of programming facility, so it is not surprising that automation nowadays tends to be associated with computer control.

Automation thus encompasses both the automatic control of events and the automatic control of variables, and this is illustrated in Figure 1.5. For example, simple camshafts or computers can be used to control the order of events, but if there is no control (or if no control is necessary) over the variables at each stage of the sequence, then the system is not an example of

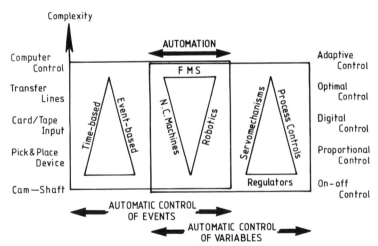

Figure 1.5 Automation encompasses control of events and control of variables.

true automation. Thus the common pick-and-place device for lifting and transferring components is excluded from the list of automated systems. And, surprisingly, the original 'Detroit automation' is also excluded. It is now classified as a transfer line. On the other hand, simple on–off control or the much more complicated adaptive controllers can be used to control the physical variables at any particular stage in a process, but if there is no control (or if no control is necessary) over the order in which the various processes are carried out then again the system does not represent true automation. Thus thermostatic control of temperature or even digital control of position does not count as a truly automated system.

True automation is represented by the overlapping area of Figure 1.5. Examples of automation abound: chemical processing, steel production, transportation, warehousing, to name but a few. However, space limitations restrict us to an examination of those examples of automation that are directly relevant to the subject of this book—robotics, numerically-controlled machines and their combination in flexible manufacturing systems.

1.6 Robotics

The robot is the epitome of automation. The word itself was first introduced in 1917 by Karel Capek in his short story 'Opilec', but it received its greatest publicity in 1920 in his famous play *RUR* (Rossum's Universal Robots): *robota* is the Czech word for drudgery or forced labour. Since then a vast number of fictional robots have appeared, many with evil intentions towards mankind, but the history of real robots did not begin until 1954 when a patent entitled 'Programmed article transfer' was filed by an American engineer called George Devol. That patent led to the first industrial robot, manufactured in 1962 by Unimation Inc., a company founded by Joseph Engleberger in 1958 (Engleberger, 1980).

Like automation, the industrial robot has attracted a multitude of definitions but the one adopted by the Robot Institute of America now receives wide acceptance: an industrial robot is a reprogrammable multi-functional manipulator designed to move material, parts, tools or specialized devices through variable programmed motions for the performance of a variety of tasks. The key words distinguishing robots from other machines are 'manipulator' and 'reprogrammable'.

Manipulation is the act of grasping an object and changing its position and orientation in space. Humans spend a lot of their time manipulating objects: the act of lifting a pen and writing with it is a typical example. In carrying out such tasks the manipulator may be required to produce up to six independent motions. This is necessary since the position of an object in space is determined by its three coordinates in a fixed orthogonal frame, and by its angular rotations around each of these three axes. In practice this means that a general-purpose manipulator requires at least six actuators

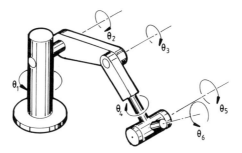

Figure 1.6   The six degrees of freedom of a manipulator.

(Figure 1.6). Many manipulators are anthropomorphic, i.e. they look like human arms; this is not surprising since they are often intended to replace human workers, and they have to fit into environments designed for humans. Thus the manipulator is often referred to as the arm, the wrist and the hand of the robot, although more recently the term 'end-effector' has been preferred to 'hand' since it encompasses tools, suction and magnetic devices as well as anthropomorphic grippers.

However, the ability to manipulate objects is not in itself sufficient to distinguish the robot from other machines. Some of the earliest automata— machines which move and act of themselves—involved manipulation. The 'child writer' of Jaquet Droz (1752–91) looked like a curly-headed four-year-old. It held a goose quill in its right hand and when the mechanism was started it dipped the pen into an inkwell, shook it twice, put its hand to the top of the page and began to write. More recently, the advent of automation has led to a multitude of automatic machines for grasping and transferring components during production processes. For example, many printing presses use grippers to pick up each sheet of paper in turn and insert it into the press. These machines have one thing in common with automata: they are designed to do one thing, and one thing only. Being inflexible and non-versatile, such systems are justified only if the particular fixed program promises to be of use over a long period of time. By definition such systems, although involving manipulation, are excluded from the industrial robot category.

There is some confusion about the relationship between industrial robots and tele-operators, especially since the latter can often look re-markably like the former (Johnsen and Corliss, 1969). For example, the earliest tele-operators, used for handling radioactive materials, looked like robot arms. But the difference is clear: the tele-operator is driven by a human operator, the robot is automatic. We shall see later (Chapter 10) that it is possible to design systems which alternate between the role of tele-operator and that of robot.

1.7 The elements of an industrial robot

In anthropomorphic terms an industrial robot requires a brain, senses, a blood supply, an arm, wrist and hand with the appropriate muscles, and possibly legs and feet, again with the associated muscles. In a typical industrial robot the equivalent machine elements could be a computer, measuring devices, electrical/hydraulic/pneumatic power, a manipulator and possibly wheels. These are illustrated in Figure 1.7.

The earlier discussion of manipulation will have clarified the need for arm, wrist and end-effector. It also argued for the need for at least six degrees of freedom for a general-purpose robot, and we shall see in Chapter 2 that there are many robot configurations capable of six degrees of freedom.

The finite lengths of the various mechanical elements restrict a robot's end-effector to a particular working volume, and it is necessary to ensure that the task to be accomplished does not call for movements beyond the boundaries of this volume. Typical working volumes are adequate for most production processes, but applications such as automated warehousing make it necessary to increase the working volume by giving the robot mobility. Future applications such as the domestic robot will also demand mobility. The obvious solution to this problem is to supply the robot with wheels, but these require a flat smooth terrain. Applications over rough terrain, envisaged by the military, and indoor requirements for stairclimbing, may require the use of legs rather than wheels. This will be examined in detail in Chapter 3.

Arms, wrists, end-effectors and legs need muscles for actuation, and in practice they may be driven by pneumatic, hydraulic or electric power. Pneumatics is restricted, on the whole, to pick-and-place robots where the actuators are allowed to move quickly until arrested by mechanical endstops.

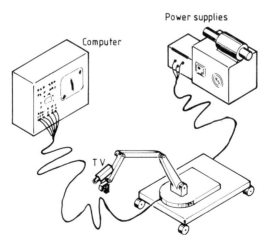

Figure 1.7 The elements of a robot system.

However, the compressibility of the air makes accurate control of speed and position extremely difficult, so that pneumatics are rarely found in the more demanding robot systems. For such systems the major contenders are hydraulic and electric drives. Hydraulic actuators are compact and capable of large forces or torques. They are presently the most popular form of power supply, particularly for larger robots. Electric drives, whilst growing in popularity with new technical advances, still only serve around 30% of the robot population. Electrically-driven robots tend to be more accurate than their hydraulic counterparts but, unlike hydraulic drives, electric motors require reduction gearboxes, and these add to the cost of the system. A more detailed comparison of hydraulics, electrics and pneumatics will be presented in Chapter 6.

Moving on to the sensing requirements of a robot, we distinguish two categories: internal and external sensing. If an end-effector is demanded to move to a particular point in space, with a particular orientation, the various mechanical elements—trunk, arm, wrist—will have to be driven to the requisite positions. Measuring devices have to be installed at each degree of freedom so that the robot knows when it has achieved those positions. This internal sensing may be carried out by potentiometers or other position/rotation measuring devices.

External sensing, on the other hand, is the mechanism of interaction with the robot's environment. First-generation robots have no such interaction, but the second-generation robots are now equipped with sensors for sight and/or touch. Robot vision, based on television techniques, will allow a robot to recognize a particular component, determine its position and orientation and then command its actuators to drive the end-effector to that position. Robots with a sense of touch, possibly derived from strain gauges, will be able to react to forces generated during automatic assembly. Sensing, both internal and external, will receive further attention in Chapters 6 and 8.

The brain or robot controller usually takes the form of a microprocessor or minicomputer. The controller has three main functions:

- to initiate and terminate motions of the manipulator in a desired sequence and at desired points.
- to store position and sequence information in memory
- to interface with the robot's environment.

We have already met some of the more elementary controllers, in the form of camshafts or plugboards applied to sequence controls. In such systems the initiation of each event is determined by the controller and the actual motion itself is controlled by mechanical stops which restrict the robot's motion between two endpoints on each axis. There is no feedback (other than information about event completion), so these systems are often referred to as non-servocontrolled robots. They are also called endpoint, bang-bang, pick-and-place or limited-sequence robots. Their open-loop nature restricts them to relatively simple tasks such as transferring parts from one place to another. It should be noted that in order for such systems to qualify as industrial robots, they must be *reprogrammable*. In practice, although the ordering of events is often easily done at a computer keyboard, the programmability of many non-servocontrolled robots is low because of the complex arrangements of endstops, limit switches and interlocks which determine the magnitude of the motions. On the other hand, they have several advantages such as low cost, accuracy, reliability and simplicity.

Non-servocontrolled robots are not suitable for applications which require an end-effector to move to a variety of positions within a working volume. In such cases it is necessary to use *servocontrolled* robots. As the term implies, these are closed-loop systems in which the controlled variables need to be measured. In addition to allowing control of position, such systems can be used to control velocity and acceleration. Feedback signals are compared with predetermined values stored in the controller's memory, and the resultant error signals direct the actuators towards their targets. The end-effector can be commanded to move or to stop anywhere within its working volume. Microprocessor- or minicomputer-based programmable controllers are often used, with large memories permitting the storage and execution of more than one program, with the ability to switch to branches and subroutines. Externally generated signals, such as those from a computer keyboard, can be used to select programs from memory.

Servocontrolled robots can be separated into two major classes—point-to-point (PTP) and continuous-path (CP) robots. The program for PTP control requires only the specification of the starting and finishing positions of a particular movement. In replaying these stored points the actuators on each axis, possibly six, are driven to their individual desired positions. The actual path followed by the end-effector is not easily predictable under PTP control unless the distance between trajectory points is made small.

There are many applications that require accurate control of the path between two points; seam welding and spray painting are typical. In such cases PTP control may be inadequate and it may be necessary to employ CP control. The CP-controlled robot may be programmed in real time by grasping its end-effector and guiding it through the required motions. By sampling on a time basis, usually 60–80 Hz, data concerning position and orientation can be stored on a tape or disc and when these data are played back the filtering properties of the robot's dynamics result in a smooth continuous motion over the desired path.

The robot's task is determined by a program which will specify the order of events and the required value of the physical variables at each event. We have already argued that the ability to change this program with ease—the programmability—is a distinguishing feature of robots. Programming can take many forms, and we have already met a few. In *manual programming* the end points of each degree of freedom are fixed by physical means such as cams or limit switches, and in *teach programming* all motion points are stored in memory by guiding the end-effector through the desired motions. But in addition to these two methods there is now a growing swing towards *off-line programming*, i.e. programming that does not require the actual robot and workpieces. Off-line programming is desirable when applications require complex motions, e.g. assembly tasks. It is also desirable in applications such as small-batch production, where it may not be possible to free a robot and make the necessary prototype parts available for teach programming. We shall see later that off-line programming is also desirable when it is necessary to link the robot to other data bases, such as the output of computer-aided design (CAD) systems. An off-line program requires a robot language for describing all the necessary operations in a suitable symbolic form. One of the most popular of these is VAL, a high-level language developed by Unimation to control and program their Unimate and Puma robots.

1.8 Why robots?

We have already spent some time enumerating the shortcomings of the human operator, and showing how mechanization and automation have come to the rescue. We can labour the point a little further with respect to robots.

- They can be stronger, allowing them to lift heavy weights and to apply large forces.
- They are tireless and can easily work around the clock, seven days a week. They do not need tea breaks or lunch breaks. They very rarely fall sick.
- They are consistent. Once taught how to do a job they are able to repeat it, practically indefinitely, with a high degree of precision. Human performance tends to deteriorate with time.
- They are well-nigh immune to their environment. They can work in very cold or very hot environments, or in areas where toxic gases or radiation are hazards. They can manipulate very hot objects. They can work in the dark.

This is an impressive list of attractions. But what do the users think? Several surveys of industrial users show industry's reasons for introducing robots. For Japanese industry in 1979 the priority list was as follows (Hasegawa, 1979):

- Labour saving   44.5%
- Improvement of working conditions   24.9%
- Increased flexibility   13.5%
- Ease of production control   8.0%
- Others   9.0%

A similar survey of German industry came up with the following list of priorities (Vicentine, 1983):

- productivity increase
- labour cost reduction
- return on investment
- improved quality
- more humane work conditions

These surveys, like many others, place a high weighting on reductions in the cost of labour. This is particularly important for two reasons: labour rates have increased by a factor of 2.5 to 4 over the last 10 years; the hourly cost of a robot, taking into account depreciation, interest, maintenance, etc., has dropped by about 15% in the same period. The increase in labour rates has largely been caused by the increasing aspirations of workers, whilst the reduction in robot rates has been due to improved technology, especially the rapid advances in microelectronics. The significance of these trends becomes obvious when it is recognized that an industrial robot can work two or even three shifts a day. This, added to potential increases in quality and savings in materials, such as in paint-spraying where less paint may be used, has led to an increased awareness of the economic benefits of introducing industrial robots.

These considerations have led to an ever growing interest in the industrial robot. A survey by the British Robot Association (BRA, 1984) showed a world population of around 98 000 robots (excluding Eastern European Countries and the USSR). The distribution was as follows:

Japan	64000	France	3380	Sweden	2400
USA	13000	Italy	2700	Belgium	860
Germany	6600	UK	2623	Spain	516

1.9   Robot applications

2623 robots were operating in the United Kingdom in 1984. The distribution of known applications was:

spot welding	471	(122)
injection moulding	412	(136)
arc welding	341	(107)
machine-tool servicing	213	(48)
assembly	199	(96)
surface coating	177	(24)
miscellaneous	175	(20)
education/research	122	(41)
handling/palletizing	102	(36)
press-tool servicing	59	(11)
grinding/fettling	43	(16)
inspection test	41	(11)
die-casting	40	(2)
glueing/sealing	23	(9)
investment casting	14	(0)

The figures in brackets (totalling 679) represent those robots that were actually installed during 1984.

There are many ways of classifying applications—historical, by method of control, by industry, by level of complexity. The authors prefer to distinguish applications in terms of the use of the robot's end-effector—is it a gripper or a tool, or could it be either?

Applications where the end-effector is usually a gripper

This category includes those applications in which the robot is used to handle products, to load/unload machines, to transfer or to reorientate components. In these cases the end-effector may have a wide variety of shapes, from mechanically-operated grippers to suction pads, but its main function is to hold the component firmly whilst it is being moved or operated upon.

The application of robots for materials handling offers potential for relieving human labour from monotonous, exhausting or hazardous work. It includes the transfer of parts between conveyors or processing lines where the parts may be heavy, hot, cold, abrasive or even radioactive. In addition it has been shown that the use of robots results in lower breakage rates when fragile objects such as television tubes have to be handled. Palletizing of large, heavy products such as engine castings or chemical products is another example of materials handling by robots. At present, when palletizing, it is necessary to present the products to the robot in an ordered fashion, and when the robot has completed the program it signals for the removal of the full pallet. Second-generation robots are now being equipped with computer vision to allow them to handle products which are presented in a disordered fashion or are not easily sorted by mechanical means.

Another important application is concerned with the servicing of machines such as presses, injection-moulding machines, die-casting machines, machine tools, etc. The first three of these involve particularly unpleasant tasks for the human operator, and this is a major reason for the introduction of robots. Die-casting, for example, is hot and dangerous. In many die-casting shops the dies are automatically lubricated and ladled with molten aluminium or zinc. Industrial robots are now being used to unload the casting, to dip it into a quenching tank and then to place it into a trimming press—all extremely unpleasant tasks for a human operator.

On the other hand, the use of industrial robots for servicing machine tools is concerned more with increasing productivity than with the elimination of unpleasant tasks. It is also an important step along the road to the fully automated factory. The greatest gains are achieved when the robot serves a number of machines, grouped in a so-called cell. In servicing such a cell the robot not only handles the materials but also ensures the optimal utilization of each machine in the cell. This requires a flow of information between the robot and the other machines. Different components require different programs, and this is where the programmability of the robot offers a great advantage. A typical layout of a machining cell is shown in Figure 1.8. Input and output magazines supply the raw material to the cell and extract the finished parts (Martins *et al.*, 1983). The robot withdraws the

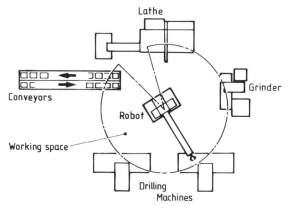

Figure 1.8   A machining cell served by a robot.

finished article from the chuck of the lathe and inserts the next with the aid of a double gripper (see Chapter 3). After machining one end of the workpiece the robot places it in a turning device which rotates through 180°: the robot then reinserts it in the chuck. The robot then presents the piece in turn to a grinding machine for surface grinding and then to a numerically-controlled (NC) drilling machine. At the completion of the machining operation the robot can present the part to a measurement station where it can check if the acceptable level of quality has been achieved. Finally the finished part is placed on an outgoing conveyor.

Applications where the end-effector is usually a tool

Spot-welding dominates this category of application. In carrying out a spot-weld the robot drives the end-effector, the weld gun, to a desired location and initiates the weld sequence of clamp, current, pressure application, current off and release. PTP control is sufficient for this task, with the robot being taught the task by a skilled welder. With first-generation 'blind' robots it is necessary to ensure that each workpiece is firmly located in the same place with respect to the robot, and this often requires the use of expensive jigs and fixtures.

The automobile industry is the biggest user of robot welders, and all major car manufacturers are involved in the new technology. Toyota's Tahara plant is a good example of the flexibility gained with robot welding: three different body styles and seven different underbodies all go down the same line at a cycle time of just under two minutes (Hartley, 1983). As each body starts down the line the production control computer switches the welding robots to the appropriate program. At present it is necessary to hold a car body stationary during the welding process, but recent developments in control systems can allow the robot to carry out the welding whilst tracking the moving car body.

Arc-welding, demanding a continuous seam of weld, requires CP rather

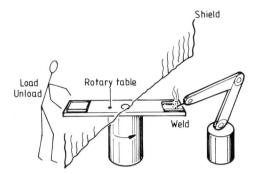

Figure 1.9 Man and robot work together on a welding task.

than PTP control. The welding head must be moved smoothly at low speed (8–25 mm/s), and the control system must ensure a path repeatability of about ±0.5 mm. Although the specification is demanding there are now more than thirty manufacturers of robots capable of arc-welding. A typical application is illustrated in Figure 1.9, which shows a welding robot being served by a rotary table (Warnecke and Schraft, 1979). An operator, positioned at the loading station, mounts the components (in this case parts of a car seat) in a jig. The rotary table then rotates through 180°, bringing the components to the welding station. The loading and welding stations are separated by a screen which protects the operator from fumes, heat and light. It appears that the productivity of robot welders is between two and four times that of manual welders, the main improvements being attributable to the robot's ability to make a long continuous welds and to make rapid between-weld movements. A major shortcoming is the inability of first-generation robots to detect wear in jigs and fixtures, springback in pressed components or heat distortion in component parts. This has led to the development of second-generation robots with the ability to track a joint, using weld current and voltage information, mechanical feelers or, more recently, vision.

Surface coating is another major application area for industrial robots. The automobile industry makes particularly heavy use of robots for painting car bodies and for spraying PVC coatings on car chasses (Hareland, 1983). The sanitary industry also uses robots for enamelling and glazing baths, basins, w.c.s, etc., and the domestic appliance industry uses robots to finish fridges, freezers, washing machines, etc. In addition to paints, other coatings such as lubricants, insulating materials, cleaning fluids, plastic resins, plasmas and chocolate have also been applied by robots. Surface coating, in particular paint-spraying, is a skilled job carried out in a most unpleasant environment. The industrial robot is able to withstand the rigours of this toxic environment and with 'lead-through' programming it is possible to transfer the skills of the human operator to the robot. Since there is a possibility of sparking when electrical drives are used, it is standard practice to use hydraulic drives in potentially explosive spraying booths.

**Applications where the end-effector can be either a gripper or
a tool**

Industrial robots are often called upon to carry out machining and finishing
operations such as drilling, grinding, fettling, cutting, routing and deburring.
The fettling of castings, for example, is a particularly unpleasant manual
task, especially the first stage which uses abrasive discs to remove runners
and risers from the casting. This task can be relegated to an industrial robot
which, with its superior strength, has the advantage over a human of being
able to handle tools of much greater power and weight. For example, in
fettling with robots, 35 kW cutting discs with 760 mm diameters are regularly
used (McCormack and Godding, 1983). In some of these machining and
finishing tasks it is more convenient and more economic for the robot to
present the workpieces to the tool rather than the reverse. Clearly this is not
feasible when the workpieces are very large, as in the fettling of heavy
castings or the drilling of holes in large aircraft panels, but it is advantageous
in situations involving a number of operations each requiring a different
tool.

The majority of finishing operations involve reactive loads, and the
resultant distortions of the robot arm are a source of inaccuracy. These can
be avoided by resorting to non-reactive processes such as laser drilling or
flame cutting. They cannot be avoided, however, in some operations such as
heavy grinding, where it is difficult to know when sufficient grinding has
taken place. Adaptive control systems are required to sense increases in
grinding pressure and to compensate automatically for offsets within the
robot program. Positioning accuracy is also of importance, particularly in
drilling and tapping, and in such cases it is often necessary to use templates
consisting of tapered drill bushings. These are usually essential when
hydraulic drives are used, but the superior accuracy of electrically-driven
robots often eliminates the requirement for expensive templates.

Another application area worthy of mention is that of inspection/test.
Robots are only beginning to be used for inspection, and there is still a lot to
be done. In manufacturing, for example, many characteristics have to be
inspected to ensure that the quality of the product is up to standard. These
include dimensions, defects and shape. Gauging devices with pneumatic,
inductive, capacitive, nucleonic or optical sensors are used to inspect
dimensions. Numerically-controlled coordinate measuring machines are also
available for rapid and precise dimensional inspection. Defects such as
cracks and flaws can be discovered by inductive, ultrasonic and X-ray
devices. Robots can be used to present such devices to the workpiece, or
vice versa. In manufacturing cells, for example, it is usual to use the robot to
present the workpieces in turn to a series of operating stations, including
inspection. In testing for leaks in car windscreens, on the other hand, robots
are trained to follow the windscreen edges with the detecting devices.

Some of the dimensional inspection devices mentioned above can also be utilized to inspect geometrical shapes or surface features, but the major developments in this area depend on computer vision. Using television/computer technology, it is now possible to teach a robot to recognize a particular shape. Hence if a product at the end of a particular process does not have a hole in the right place, its shape will not match the stored template and the robot can be taught to react by directing it to the reject bin.

The last, and perhaps the most challenging, application to be discussed in this section is that of assembly. When products have to be assembled, whether they be electric motors, carburettors or printed circuit boards, there are three options available: manual assembly, dedicated machinery, or robots. We argued earlier that dedicated machinery (hard automation) only pays off when production runs are large, and it is interesting to note that only around 6% of assembled products fall into this category. Thus assembly is a labour-intensive area of production; one that plays an important part in determining the final cost of a product. Indeed the cost of assembly may be as high as 50% of the total manufacturing cost (Crossley and Lo, 1981). To date, however, relatively few robots have been used in assembly, basically because of the variety of tasks involved, and because of the need for high accuracy. There is also the difficulty of competing with the extraordinary dexterity of the human operator who can pick a component from an untidy heap, align it properly, insert it and fix it in position. These tasks present a difficult challenge to the robot.

Fortunately the large majority of assembly tasks are relatively simple, involving vertical insertion, and screwing or pressing. Early work at the Charles Stark Draper Laboratory at MIT found that 60% of parts were inserted from one direction, 20% from the opposite direction, 10% at right angles to these directions and 10% in other directions. Thus the main operation in assembly is to pick a component up with a vertical movement, move it horizontally and then move it downwards vertically for insertion. This can be achieved by a robot with three degree of freedom, such as the one shown in Figure 1.10.

Because of the possibility of error in alignment it is necessary to build a

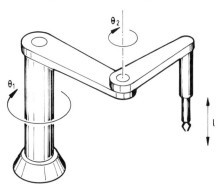

Figure 1.10   SCARA-type configuration is useful for assembly tasks.

degree of compliance into the assembly robot. In the SCARA robot, whose geometry is similar to that shown in Figure 1.10, this compliance is inbuilt since the manipulator is driven by DC motors whose braking torques can be overcome by torques arising from misalignment. This is a beneficial feature unique to the particular geometry of the SCARA. When other geometries are used it is necessary to build the compliance into the wrist. The presence of this compliance allows the robot to respond in a passive way to reaction forces, but there are now several systems that can respond actively: i.e., by measuring the forces they can make the manipulators take corrective action to ensure that they are not excessive. In addition to this tactile sense these second-generation assembly robots are being equipped with vision to allow them to carry out difficult tasks such as picking items from bins and assembling printed circuit boards. The provision of vision eliminates the ancillary equipment required to orientate and position components for first-generation robots.

Towards the automated factory

Automated assembly is a step along the road towards the automated factory in which the three major functions—design, production and administration—are integrated through the computer (Gunn, 1982). The aim is to organize, to schedule and to manage the total manufacturing enterprise, from product design to manufacture, distribution and servicing in the field. The computer's ability to store data, and to communicate data to and from other stations in the factory, makes it a central feature of the automated factory. For example, the manufacturing chain, illustrated in a simplified form in Figure 1.11, is now often augmented by computer-aided design

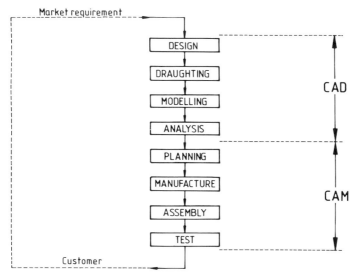

Figure 1.11 The total manufacturing enterprise.

(CAD) and computer-aided manufacture (CAM). In the widest sense CAD refers to any application of a computer to design, draughting or modelling—usually involving some form of interactive use of computers and graphic displays (CADCAM, 1982, 1983). The power of the computer can also be used for analysis, e.g. finite-element analysis and kinematic analysis. The use of the computer allows information to be stored in a database where it can be used to determine what materials need to be purchased, what parts need to be withdrawn from store and what instructions need to be given to the production machines. The stored geometric data can also be used in other programs such as the finite-analysis package, the technical illustration package and simulation packages. These data provide the bridge between CAD and CAM, which is concerned with the application of computers to manufacturing processes, for the control of machinery, for handling the data that describes the manufactured products and for the control and handling of materials flowing through the system. The use of numerical computer-based data eliminates the need for draughting, a task that could take up to 80% of design time.

CAM has progressed steadily from the early NC machines whose actions were programmed by tape. Their introduction was part of the drive to increase productivity—a drive spurred on by the recognition that in a typical stand-alone manually operated machine, only about 1.5% of the machine's time was used for productive cutting of metal (Carter, 1972). The rest of the time was taken up in tool-changing, loading, gauging, maintenance and just lying idle. NC machines were able to produce five times as much as their older manually-controlled counterparts, but even greater productivity increases were promised by computerized numerical control (CNC), in which the paper tape inputs were replaced by programs stored in a dedicated computer. Such a system had greater flexibility than NC, since programs could be easily edited and could indeed be tested before actual use by simulating on a visual display unit (VDU). In addition, programs related to different products can be called up in random order hence allowing the machine to handle different batches with minimum loss of time in changing over and setting up. This ease of programming makes the CNC machine equivalent to a robot; one especially designed for handling tools.

The next evolutionary step was the development of direct numerical control (DNC) by which a single supervisory computer was able to control a number of machines. This allowed the dedicated computers of the CNC machines to be replaced by a larger computer that managed many machines on a time-shared basis. In the simplest DNC systems up to a few dozen machine tools are run on a completely independent basis, and loading and unloading of each machine is carried out manually. Flexible DNC is more complex, the machine tools and an automated materials-handling system being connected to and related through a control computer. Such a system can process items having various shapes and requiring different machining processes. The manufacturing cell discussed earlier (Figure 1.8) was an example of such a system in which the materials-handling function was carried out by a robot.

To all intents and purposes flexible DNC systems can be taken to be synonymous with flexible manufacturing systems (FMS), the latest development in manufacture. An FMS may be defined (Ranky, 1983) as a system dealing with high-level distributed data processing and automated flow of material using computer-controlled machines, assembly cells, industrial robots, inspection machines and so on, together with computer-integrated materials handling and storage systems. A schematic FMS layout is shown in Figure 1.12 (CADCAM, 1983). The key to such systems is their flexibility. Different products require different machining operations, and these in turn require the facility to change the direction of the workflow as required. We have already seen how the robot can act as a transporting and mechanical handling device, but in large systems it is also often necessary to use automated guided vehicles (AGV), guided by buried cables or radio control. The similarity of the control systems of NC machines, robots and AGV allows the supervision of the whole operation by a central computer.

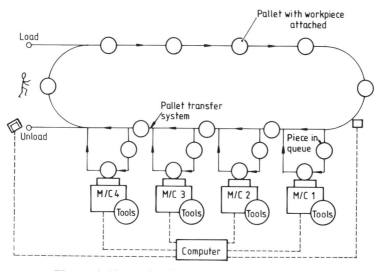

Figure 1.12   A flexible manufacturing system.

There are currently about 80 FMS in commercial application worldwide. One of the most exciting belongs to Fanuc, one of the largest Japanese manufacturers of industrial machinery, including lathes and robots (Hartley, 1983). The factory includes an automated warehouse, a machine shop, a welding shop, an inspection shop and an assembly shop. Robot carts following buried wires transfer the workpieces from the store, through the shops and to the warehouse. The total system is complex, the machine shop alone containing 29 cells, 7 using CNC lathes for processing cylindrical parts and 22 using CNC machining centres to shape prismatic parts. Robots are used for transferring workpieces between lathes and pallets.

1.10   Recapitulation

This chapter has attempted to trace the origins and evolution of the industrial robot, the epitome of automation. Starting from relatively sluggish machines, applied to boring unpleasant tasks, there has been a move towards a new generation of much faster machines, having sensory interaction with their environment and capable of carrying out delicate precision tasks such as assembly. With the introduction of off-line programming the robot no longer needs to be viewed in isolation, and its ability to communicate with other computer-based systems, such as CAD, has resulted in a growing number of integrated systems, from machining cells to FMS. The next evolutionary step, the third generation, will undoubtedly endow the robot with a degree of intelligence—artificial intelligence—so that it will become more autonomous, with the ability to learn from experience and to react to unforeseen circumstances.

Chapter 2

Mechanisms and robot configurations

2.1 Introduction

This chapter is concerned with the skeletal structure of the robot. It will show how several different trunk/arm arrangements can be used to place a wrist at a point in space, and how the wrist in turn can place the end-effector at a particular angular orientation. Trigonometric and matrix methods will be used to formulate relationships between the movements of the robot's joints and the position and orientation of its end effector.

2.2 Mechanisms

A mechanism is a means of transmitting, controlling or constraining relative movement. In reality there is always some degree of flexibility inherent in each of the constituent bodies, or links within a mechanism, but this can often be neglected allowing the mechanism to be treated as an assembly of rigid links connected to each other by joints that permit relative movement.

Figure 2.1(a) shows a planar mechanism in which the motion of the links is restricted to a plane; in this case the plane of the paper. The mechanism appears to have only three links, but there are in fact four, the fourth one being fixed to earth. The sketch in Figure 2.1(b) is an example of another class of mechanism, the spatial mechanism, whose various links can move in different directions in space. The example shown again has four links, with one fixed to earth.

The joints connecting the links are known as *kinematic pairs*. As the name implies, a pair has two elements, one attached to each of the two links

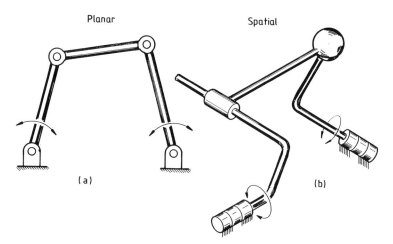

Figure 2.1   Mechanisms: (a) planar (b) spatial.

it connects. If two mating elements are in surface contact, they are said to form a *lower pair*: if the contact is at a point or along a line the pair is known as a *higher pair*. For example, all the joints in Figure 2.1 are lower pairs. Higher pairs are typified by gears and cams. A mechanism all of whose joints are lower pairs is commonly called a *linkage*: hence Figure 2.1 illustrates two types of linkages.

Space does not permit a full discussion of pairs, so we shall have to restrict outselves to those lower pairs shown in Figure 2.2. Pair (a) is a turning or revolute pair R. It allows only one relative movement between the two links—a rotation about the axis of the pair. For this reason it is said to have one degree of freedom; $f = 1$. Figure 2.2(b) shows a sliding or prismatic pair P. The elements of a P pair are congruent prisms or non-circular cylinders. It has one translatory degree of freedom; $f = 1$. P pairs and R pairs can be used in both planar and spatial mechanisms.

The remaining lower pairs in Figure 2.2 are of the spatial variety. Figure 2.2(c) shows a cylindrical or C pair, consisting of two identical circular cylinders; one convex, the other concave. The elements can rotate relative to each other about the common axis of the cylinders: they can also translate relative to each other along this axis. Since these two motions can exist entirely independently of each other the C pair is said to have two degrees of freedom; $f = 2$. Figure 2.2(d) shows another lower pair with two degrees of freedom, but in this case both freedoms are rotary ones. The slotted spherical pair or SL pair consists of a convex or solid sphere which, ideally, exactly conforms with a spherical shell of identical radius. The pin and slot stops relative rotation about the vertical axis. In Figure 2.2(e), on the other hand, there is no such restriction and therefore the ball joint or spherical S pair has three rotary degrees of freedom; $f = 3$.

There are other types of pairs, such as screw pairs and planar pairs, but

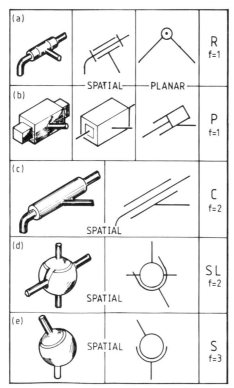

Figure 2.2 Pairs: (a) revolute (b) prismatic (c) cylindrical (d) slotted spherical (e) spherical.

Figure 2.2 provides us with a variety sufficient for the purposes of this book. Our objective is to demonstrate how pairs affect the degrees of freedom of a mechanism, and this in turn will allow us to propose a variety of possible robot configurations.

Number synthesis

We have just met the concept of degrees of freedom as applied to pairs. Another way of looking at it is to define the number of degrees of freedom of a pair as the number of independent variables which must be specified in order to locate its elements relative to one another. This concept can be extended to a consideration of the complete mechanism. The degrees of freedom of a mechanism are often referred to as its *mobility*. Analogous to the freedoms of a pair, mobility is defined as the number of independent variables which must be specified in order to locate all the members of the mechanism relative to one another (Hunt, 1978). Referring back to Figure 2.1(a), a little bit of thought will confirm that $M = 1$; i.e. if any one angle between the constituent links is known or fixed, then all other angles will be known or fixed.

This technique, known as number synthesis, allows us to determine the mobility of most mechanisms (there are some special cases that require care). In applying the technique we have first to recognize that a free rigid body has six degrees of freedom: it can move along any of three orthonormal axes and it can rotate about each of them. In other words we have to know three linear displacements and three angular rotations before we can define its position in space. Hence N unconstrained bodies will have $6N$ degrees of freedom. Now in order to convert these bodies into a mechanism it is necessary to fix one to earth, and to connect them by pairs. Both of these operations reduce the total number of degrees of freedom, or mobility, of the mechanism. In fixing one link, six degrees of freedom are removed: hence the mobility M is reduced to $6(N-1)$. In addition, each pair reduces the mobility by $(6-f)$ where f is the number of degrees of freedom of the particular pair. For example the S pair has $f = 3$, hence when it is used to connect two links it removes three degrees of freedom from the connected links; i.e., one of them has lost its three degrees of translatory freedom. Thus, with G pairs in total, the mobility M becomes

$$M = 6(N-1) - \sum_{i=1}^{G} (6 - f_i) \tag{2.1}$$

where f_i is the number of degrees of freedom of the ith pair, or

$$M = 6(N - G - 1) + \sum_{i=1}^{G} f_i \tag{2.2}$$

This equation is often attributed to Grübler.

Let us apply it to the spatial linkage of Figure 2.1(b). Here $N = 4$ and $G = 4$. There are two R pairs, one C pair and one S pair, giving a total freedom at the joints of $2 + 2 + 3 = 7$. Hence

$$M = 6(4 - 4 - 1) + 7 = 1$$

In the case of planar mechanisms, such as that of Figure 2.1(a), the above equation must be modified to take into account the fact that a rigid body has three, not six, degrees of freedom in planar motion. The reader should verify that the modified equation becomes

$$M = 3(N - G - 1) + \sum_{i=1}^{G} f_i \tag{2.3}$$

Let us start by considering the linkages shown in Figure 2.3. The first row shows three linkages each having zero mobility. For example, linkage (b) has $N = 5$ and $G = 6$. Note that one of the links has three pairs attached to it: for this reason it is known as a ternary link. The rest are binary links. For each R pair $f = 1$. Hence

$$M = 3(5 - 6 - 1) + 6 = 0$$

Such arrangements cannot be used as mechanisms, because there cannot be relative movement between any of the links. They are in fact statically determinate structures.

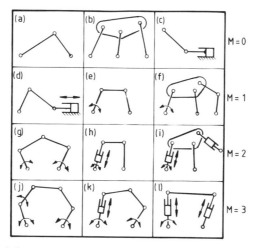

Figure 2.3 Linkages with mobilities $M = 0$, 1, 2 and 3.

The second row of Figure 2.3 shows a more useful group of linkages. Here $M = 1$ and it is possible to identify an input which, being known, determines the value of an output. For example in linkage (d), employing three R pairs and one P pair, the input could be the displacement of the P pair and the output could be the resultant rotation of the left-hand link. We have already met linkage (e) in Figure 2.1(a). Linkage (f) has been included to demonstrate that the addition of the right-hand pair of links to linkage (e) does not affect the overall mobility. This pair of links, with its three R pairs, is known as a *dyad*. A quick calculation confirms that its introduction does not affect the mobility—there are two additional links, adding six degrees of freedom, but there are three additional R pairs removing six degrees of freedom.

Linkages (g), (h) and (i) each have $M = 2$. For example, (i) has $M = 7$ and $G = 8$, including two P pairs and six R pairs. Hence

$$M = 3(7 - 8 - 1) + 8 = 2$$

For each of these three cases an output depends on two inputs. Note also another form of dyad, shown in linkage (i), consisting of two links with two R pairs and one P pair. Its addition to (h) does not change its mobility.

Finally, the last row shows linkages with $M = 3$. Linkage (k) has been generated from (j) by replacing the left-hand RRR dyad by an RPR dyad. Linkage (l) replaces the remaining RRR dyad by another RPR dyad. So dyads can be introduced or exchanged without affecting a mechanism's mobility.

Coupler curves

Now this may be all very well, but what has it got to do with industrial robots? The answer is that mechanisms are designed so that some particular

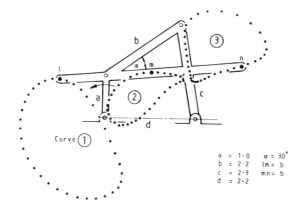

Figure 2.4   Coupler curves for a four-bar linkage.

point or link moves in a known way in response to known input commands. This is just what we expect of the end-effector of an industrial robot, which has to move in a particular way in response to given inputs.

In the kinematics of mechanisms the study of the motion of a particular point on a link falls within the theory of coupler curves (Hunt, 1978). The hinged planar four-bar linkage is often used to trace certain paths. It was shown in its most elementary form in Figure 2.1(a), but it should be noted that each of the three moving links could be extended indefinitely along and on either side of its centre line. Thus for each particular geometrical arrangement of the pairs, there is an infinite number of designs for four-bar linkages. Figure 2.4 shows a particular four-bar chain with links *a, b, c* and *d*. Link *b,* the coupler, has been extended, and the figure shows the paths, or coupler curves, traced out by three different points, *l, m* and *n* on the coupler, as the link *a* rotates at constant velocity. (Note that links *a* and *c* have to be offset to allow *b* to rotate.) The distance between successive points is a measure of the velocity; wide spacing indicates high velocity.

A knowledge of such curves is immensely useful to the designer of mechanisms. For example, curve (3) has a straight-line portion and could therefore be used for feeding components to a machine, or pulling film through a projector. Curve (1) imitates to some extent the motion of the human foot and leg, and we shall see later (Chapter 3) that this allows it to be used in walking machines. Large atlases have been produced of coupler curves, for different points on hundreds of planar linkages (Hrones and Nelson, 1957).

Robot mechanisms with *M* = 1

The four-bar linkage of Figure 2.4 has $M = 1$ so there is a unique relationship between the position of a point on the coupler and the position of the input link. The fact that the coupler point is constrained to move along a particular path reflects the one-degree-of-freedom nature of the

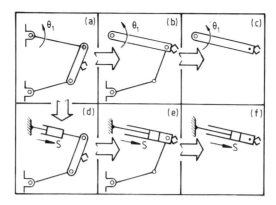

Figure 2.5   Six different one-degree-of-freedom manipulators.

linkage. Figure 2.5 illustrates how this property could be used in elementary robots. Six different configurations are shown, each with $M = 1$, and each with an end-effector attached. In cases (a), (b) and (c) the position of the end-effector is uniquely determined by an angular input θ. In (d), (e) and (f) the input is a translation S.

There is a logical development from the four-bar linkage of (a) through to the one-link open linkage at (c). First, the end-effector is moved from the middle of the coupler, where it would traverse quite a complicated path, to the end of the input link where it traverses an arc, or possibly the circumference, of a circle. Configuration (c) is then obtained by removing the dyad from (b). Now (c) is unusual in the sense that, unlike the others, it is an open linkage. It illustrates the fact that the concept of mobility applies to open as well as closed linkages. Equation (2.3) could not, however, be applied directly to the open linkage, but its mobility could have been deduced by noting that it was kinematically equivalent to (b), or by simply subtracting the two constraints of the single R pair from the three planar degrees of freedom of the single link. Now this observation is not merely pedantic: it will be seen later that, whereas (c) may often be the orthodox choice for a robot arm, there can be advantage in using (a) or (b) instead.

Configurations (d), (e) and (f) are also logically related. Linkage (d) is obtained from (a) by replacing an R pair in (a) by a P pair. The input in this case is a translation S. When the dyad is removed from (e) we are left with another elementary open linkage with one degree of freedom. This time, unlike (c), the freedom is a translatory one, and the end-effector moves in a straight line.

Robot mechanisms with $M = 2$

A robot with only one degree of freedom would be seriously restricted: let us move on to look at mechanisms with $M = 2$. Ten variations are shown in Figure 2.6, each allowing the end-effector to move to any point within a

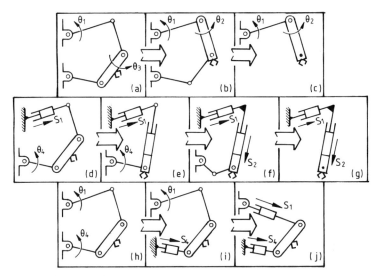

Figure 2.6 Two-degree-of-freedom manipulators.

limited two-dimensional space. Consider first the configurations (a), (b) and (c). In (a), since $M = 2$, two inputs θ_1 and θ_3 are used to position the end-effector. In (b) θ_1 and θ_2 are used as inputs and (c) is derived from (b) by removing the lower dyad. The two-link open linkage of (c) is the orthodox serial-operated robot arm.

Linkage (d) is derived from (a) by replacing an R pair by a P pair. If the right-hand RRR dyad in (d) is then replaced by an RPR dyad the configuration in (e) results. Configuration (f) is achieved by swopping freedoms within linkage (e): an R pair is moved from the top to the bottom. The resultant RRR dyad can then be removed to give the two-link open chain of (g) which, like (c), is an orthodox means of obtaining two degrees of freedom in a robot, although the two translational freedoms are normally chosen to be at right angles.

Finally (h), which is similar to (a) except that θ_4 is used as an input instead of θ_3, changes to (i) when an R pair is replaced by a P pair. Exchanging the RRR dyad in (i) for an RPR dyad then gives configuration (j), which like (g) has two translatory inputs, but this time in parallel rather than in series. This mode of operation can have several advantages: for instance, it provides a more rigid structure than the serial-operated configuration and this leads to improved accuracy of positioning. Also, the actuators can be positioned close to the frame; they do not have to be carried around as in (c) and (g). Hence the whole configuration can be relatively lightweight, leading to increased speed of response.

Figure 2.6 is not an exhaustive catalogue of means of guiding an end-effector in a two-dimensional space. There have been several recent developments, some of them based on the pantograph. The pantograph linkage is used to magnify or reduce a figure drawn on a plane. Figure 2.7(a)

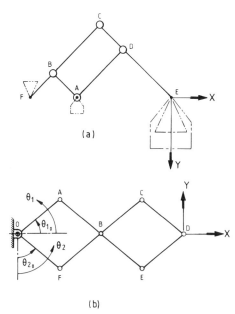

Figure 2.7 Pantograph linkages.

shows one possible arrangement with $FB = BA$ and $BC = AD = DE$. Here, if A is a fixed point, point E will invert and magnify the motion of point F by an amount BC/BF. If, on the other hand, F is used as the fixed point and A is free to move then E will magnify the motion of A by an amount $(1 - BC/BF)$. This linkage allows linear actuators to be used for positioning point E in the plane X, Y. If a vertical actuator is situated at F and a horizontal one at A, then for a desired position x, y of E, actuator F would need to move F upwards by $y/(BC/BF)$ and actuator A would need to move A to the right by $x/(1 + BC/BF)$. We shall see later that control in cartesian coordinates is easier to achieve than other methods, such as the jointed arm of Figure 2.6(c). Figure 2.6(g) showed another way of achieving control in cartesian coordinates, but whereas Figure 2.7(a) has actuators fixed to ground at A and F, one of the actuators in Figure 2.6(g) has to be carried around, thereby increasing the moving mass and reducing speed of response. The pantograph linkage of Figure 2.7(a) is used in commercial robots (Hirone and Umetani, 1981) and in walking machines (Taguchi *et al.*, 1976).

Another pantograph configuration is shown in Figure 2.7(b). It consists of six links with $OA = OF = CD = DE$ and $AE = FC = 2OA$. In this case two rotational inputs θ_1 and θ_2, operating in parallel, are used to position point D in the plane X, Y, rather than the linear inputs of Figure 2.7(a). The movement of point D is twice that of point B, whilst the movement of B is the sum of the movements of points A and F. It is the latter fact that makes this such a useful linkage. If the initial values of θ_1 and θ_2 are θ_{10} and θ_{20}, defined in Figure 2.7(b), it is easy to show that the movement of D is given

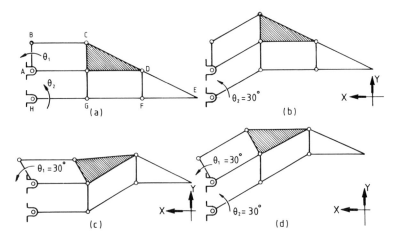

Figure 2.8   The parallel linkage manipulator: (a) initial configuration (b) the effect of a 30° change in θ_2 (c) the effect of a 30° change in θ_1 (d) the combined effect.

by

$$x = 2(OA)(\sin\theta_2 + \cos\theta_1 - (\sin\theta_{20} + \cos\theta_{10})) \qquad (2.4)$$

and

$$y = 2(OA)(\sin\theta_1 - \cos\theta_2 - (\sin\theta_{10} - \cos\theta_{20})) \qquad (2.5)$$

Another linkage with two degrees of freedom is shown in Figure 2.8. This is described in Toyama and Takano, (1981) as a parallel linkage type. In the case shown the links BC, AJ, HG, GF and FE are of equal length and $AB = CJ = AH = DF = 0.5HG$. As in Figure 2.7(b) two parallel rotary inputs θ_1 and θ_2 are used to control the X, Y-coordinates of point E. Figure 2.8(b) shows what happens when HG is rotated through 30° whilst keeping AB fixed ($\theta_1 = 0$). You will note that the movement of E is similar to that of G. In Figure 2.8(c) HG is fixed ($\theta_2 = 0$) and AB is rotated through 30°. In this case the movement of E is one half of that of B. Figure 2.8(d) illustrates the result of combined rotations, $\theta_1 = \theta_2 = 30°$ and it shows that the movement of E is the sum of the movements in Figure 2.8(b) and (c). This particular configuration was developed in an attempt to speed up the motion of robot arms. The argument is that it has higher rigidity than conventional robot arms and that the torques required to hold a stationary load are minimal.

Robot mechanisms with $M > 3$

We now move on to systems with three degrees of freedom, and we shall continue for the time being to restrict ourselves to planar motion. The astute reader, however, will notice that all of the systems with two degrees of

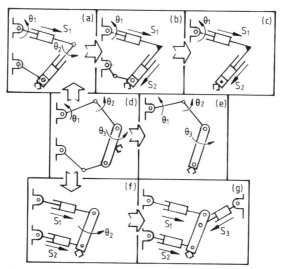

Figure 2.9   Mechanisms with mobility $M = 3$.

freedom discussed so far (Figures 2.6, 2.7, 2.8) could have been given another degree of freedom by allowing rotation about a vertical axis.

Figure 2.9 shows seven linkages, each with a mobility of 3. Starting with (d), which we have already met in Figure 2.3(j), we note that with three inputs θ_1, θ_2 and θ_3 we are now able to control both the position of the end-effector and its orientation. The closed linkage of (d) can be converted to the more conventional open linkage (e) by removing the bottom RRR dyad.

Linkage (a) can also be derived from (d), but in this case two RRR dyads are replaced by RPR dyads. Linkage (b) is obtained by transferring freedoms within (a). Linkage (c) is obtained by removing the RRR dyad from (b).

Another variant of linkage (d) can be achieved by replacing the top and bottom RRR dyads by RPR dyads as in (f). Finally (g) is obtained by adding an RPR dyad to (f). This allows all inputs to be translational and in parallel, compared to the serial nature of the open linkages (c) and (e).

We mentioned earlier that parallel operation offers several advantages, such as rigidity and speed of response. This has led to a considerable interest in parallel systems, especially for assembly operations. Many of the proposed configurations spring from earlier work by Gough (1962) and Stewart (1965). Both of these authors designed and built platforms with six degrees of freedom, in Gough's case for testing tyres and in Stewart's case for aircraft simulation.

In order to understand the operation of such systems, let us start by examining a system with three degrees of freedom. The linkage of Figure 2.9(g) is a starting point: it is shown in a more symmetrical arrangement in Figure 2.10(a). If the configuration shown is taken to be the reference

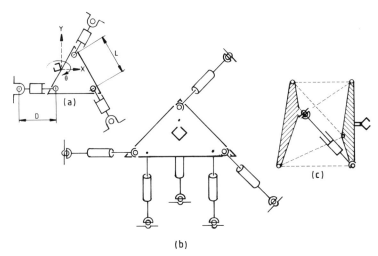

Figure 2.10 Parallel topology: (a) three degrees of freedom (b) six degrees of freedom (c) a compact version of (b).

position, then, for example, the required extension of the left-hand actuator for a given position (x, y) and orientation θ of the end effector is

$$\{[x + D + 0.25L - 0.5L \cos(60 - \theta)]^2 + [y + 0.433L - 0.5L \sin(60 - \theta)]^2\}^{0.5} - D \quad (2.6)$$

Similar expressions can be developed for the extensions required of the other two actuators. We shall return to this equation later when we have discussed techniques for determining actuator motions for serial-operated systems.

The mobility of this system can be increased from 3 to 6 by introducing another three linear actuators (or prismatic pairs), one placed vertically at each corner of the platform (Figure 2.10(b)). We now have a spatial linkage allowing the end-effector to be positioned at x, y, z and with specified orientations with respect to each of the X-, Y- and Z-axes. Spherical pairs S are used to connect the actuators to the platform, whilst SL pairs (two-axis joints) are used to connect the actuators to ground. (The alignment of these SL pairs needs careful consideration.)

Mobility can be checked by using eqn (2.2):

$$M = 6(N - G - 1) + \sum_{i=1}^{G} f_i$$

Here $N = 14$ and $G = 18$. Of the 18 lower pairs, there are six S pairs each with $f = 3$, six SL pairs each with $f = 2$, and six P pairs each with $f = 1$. Hence

$$M = 6(14 - 18 - 1) + (6 \times 3) + (6 \times 2) + (6 \times 1)$$

$$= -30 + 36$$

$$= 6$$

More compact versions of this configuration have been proposed (Hunt, 1978; 1982) one being shown in Figure 2.10(c) where, for the sake of clarity, only one of the six linear actuators is shown.

Here six degrees of freedom were achieved by moving the platform as a whole. However, in the majority of conventional robots the six freedoms are split between the robot's arm and its hand. Basically the arm positions the wrist in three-dimensional space and the wrist in turn determines the orientation of the hand with respect to three orthonormal axes.

Robot mechanisms with $M > 6$

An interesting situation arises when the number of degrees of freedom of a robot mechanism (its mobility) exceeds the maximum number of 6 required by an end-effector. What use are mechanisms with such high mobilities? The answer is obvious: they are highly mobile. They can therefore reach into corners that would be inaccessible to robots with lower mobilities.

However, when $M > 6$ there are too many unknowns in the kinematic equations and a problem of mathematical redundancy arises. This can be explained by reference to the simpler case of $M = 2$ shown in Figure 2.11(a), where the end-effector needs two degrees of freedom so that it can be positioned at any point in the XY-plane. The open linkage of Figure 2.11(a) has $M = 2$ and can therefore provide these degrees of freedom. Two equations can be set up relating θ_1 and θ_2 to x and y, and given x and y, these can be solved for θ_1 and θ_2.

The open linkage of Figure 2.11(b) has $M = 4$, but although there are now four variables the number of equations remains at two; one for x and one for y. Hence for a given x and y there is an infinite number of solutions for the angles θ_1 to θ_4. This intractability can be avoided, and some of the

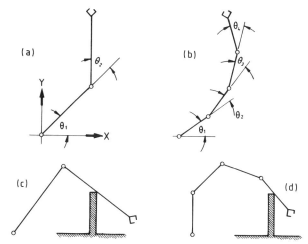

Figure 2.11 Manoeuvrability: (a) two degrees of freedom (b) dummy freedoms (c) and (d) manoeuvrability is increased.

advantages of high mobility can be maintained, by introducing the concept
of dummy freedoms. For example, if θ_2 were constrained to equal $k\theta_1$ and
θ_4 to equal $K\theta_3$, then θ_2 and θ_4 would be dummy freedoms. Their use,
whilst not increasing mobility, can give rise to a useful increase in a robot's
manoeuvrability. This is illustrated in Figure 2.11(c) and (d) which show a
robot arm attempting to reach over an obstacle. In (c) the arm has $M = 2$. In
(d) each link is half the length of that in (c), but dummy freedoms have been
introduced at the second and fourth joints. Here $\theta_2 = 0.5\theta_1$ and $\theta_4 = 0.5\theta_3$.
Clearly (d) is more manoeuvrable than (c).

2.3   Simple chains: $M = 3$

Figure 2.6, 2.7 and 2.8 show 13 ways of achieving two degrees of freedom:
there are many others. However, we should note that any of these
configurations could be converted to a three-degrees-of-freedom system by
allowing rotation about a vertical axis. For example, the system in Figure
2.6(c), which is anthopomorphic in that it resembles the human arm, can be
described as an RR system since it employs two revolute joints. Freedom to
rotate about a vertical axis, or trunk, would convert this to an RRR system
like the one shown in Figure 2.12, variant 9. Figure 2.12 is a compilation by
Milenkovic and Huang (1983) of the major three-degrees-of-freedom robot

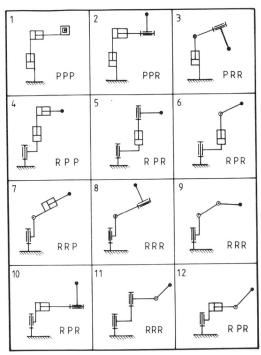

Figure 2.12   The major three degrees of freedom robot linkages (Milenkovic and
Huang 1983).

linkages. It does not include any of the closed linkages of Figure 2.6. It restricts itself to simple chains, defined as open linkages which use only rotary or prismatic pairs, with the pair axes either perpendicular or parallel to each other.

In all there are 36 possible combinations of these simple three-pair chains. Of these, nine degenerate into only one or two degrees of freedom, e.g. three P pairs parallel to each other can only produce translation in one direction. Of the remaining 27 possibilities, seven are planar, e.g. three R pairs parallel to each other can produce only planar motion (See Figure 2.9(e).) This leaves 20 possible spatial, simple chains. But even these can be further refined, for there are certain configurations in which the order of the pairs does not affect the motion of the arm. For example, consider variant 4 in Figure 2.12. This is known as a cylindrical robot. No change in the motion of the arm results if either the vertical P pair precedes the R pair or the vertical P pair follows the horizontal one. An elimination of all of these equivalent linkages reduces the total number to 12, as shown in Figure 2.12. Where equivalents exist, the representative of the group is chosen to be the one in which the P pair that is parallel to an R pair is in the middle, as in variant 4. Can you spot the others?

This offers the designer a wide variety of possibilities. To date however, only four of the configurations of Figure 2.12 have found widespread use. Variant 1 is known as the *XYZ* or *cartesian* configuration; variant 4 as the *cylindrical* configuration; variant 7 the *polar* or *spherical* configuration and variant 9 the *jointed-arm* or *revolute* configuration. Figure 2.13 will make the reason for these names clearer.

Figure 2.13(a) shows a side elevation and plan view of a cartesian robot.

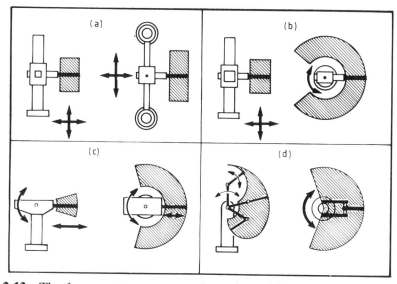

Figure 2.13   The four most common configurations: (a) cartesian (b) cylindrical (c) spherical (d) jointed arm.

All three actuators are linear and movements take place along the X-, Y- and Z-axes. The shaded area illustrates the extent of the working envelope, a cuboid in this case. A cylindrical robot is illustrated in Figure 2.13(b). The working envelope is a portion of a cylinder and any point in the envelope is defined in terms of its cylindrical coordinates, i.e. an axial position, a radial position and an angle. Figure 2.13(c), on the other hand, shows a robot which operates in polar coordinates. Its work envelope is a portion of a sphere (hence its common name), and points within the envelope are defined in polar coordinates, i.e. a radial position and two angles. Finally, Figure 2.13(d) is a sketch of a jointed-arm robot, an anthropomorphic configuration reminiscent of the human trunk and arm. Here again the working envelope is a portion of a sphere.

It is difficult to generalize when comparing these different configurations: their choice is often dictated by the application. However, on the whole, it can be said that the work envelope improves as we move from (a) to (d). In (d) for example, it is possible for the end-effector to be tucked in close to the base as well as working at a distance. This allows the jointed arm to reach into bins and over obstacles. Ease of control however, tends to go the other way, being best for (a) and worst for (d). In the cartesian robot, for example, movements can take place along any axis without disturbing the angular orientation of the end-effector. This is not so for the spherical robot, where rotation about the vertical Z-axis would change the orientation of the end-effector with respect to the X- and Y-axes. There could be occasions where this would be unacceptable and the control circuitry would have to provide compensatory movements of the end-effector's motors. This problem is even more severe for the polar configuration, with its two rotations. It is worst of all in the case of the jointed arm, where it is compounded by the fact that even a straight-line movement along a single axis requires a coordinated motion of elbow and shoulder joints.

The accuracy of the robot can also be affected by its configuration. We shall see later that most robots operate as closed-loop systems—they compare the actual positions of the various control coordinates with the desired values and take corrective action when an error is detected. Now the actual position is determined by measurement, and the ultimate accuracy of the system depends on the accuracy of this measurement. The resolution of a control system is defined as the smallest change in position that the feedback device can sense. For example, an optical encoder is often used for position measurement in a machine tool. Such an encoder, emitting 1000 voltage pulses per revolution, could be attached to a 10 mm pitch leadscrew on a machine-tool table and the encoder would then emit one pulse for each $10/1000 = 0.01$ mm of table displacement. Thus the control resolution would be 0.01 mm.

We should be able to see that, because of its linear movements, the cartesian configuration is capable of a resolution of the same order as that regarded as the state of the art in machine tools (0.01 mm), and this resolution could be maintained throughout the whole working volume.

This latter property cannot be maintained when a configuration employs a rotary motion. Consider the cylindrical robot of Figure 2.13(b) and imagine that an optical encoder were used to determine the angular rotation about the vertical trunk. If this encoder emitted 1000 pulses/rev then the angular resolution would be $360/1000 = 0.36°$. This would appear to offer reasonable accuracy if we were only attempting to control the angular rotation. But the objective is to control the position of the end of the arm, and if it were typically 1 m long the positional resolution of its tip would be $1000 \times 0.36 \times \pi/180 = 6.28$ mm. This is many times greater than the value expected from machine tools.

A similar problem arises in the spherical robot, where, whilst the linear movement can be controlled accurately, positional changes caused by angular motions can have poor resolutions.

The jointed-arm robot, with its three rotary movements, is worst of all in this respect. And, in addition, its articulated structure allows the joint errors to accumulate.

Thus, in summary, complexity of control tends to increase in going from robot (a) to robot (d). Problems of accuracy, due to flexibility and poor resolution, also tend to increase. On the other hand, speed of movement of the end-effector tends to increase whilst going from (a) to (d) and the size of the working volume and its accessibility gets better.

2.4 Geometry of simple chains

The geometries of the basic configurations are shown in more detail in Figure 2.14. (The cartesian robot is not shown because of its simplicity.) For the cylindrical robot H and z are the linear inputs and ϕ the rotational one.

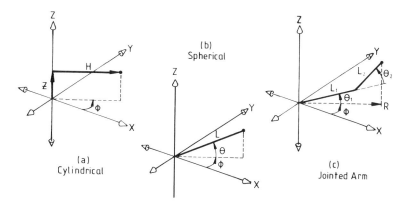

Figure 2.14 Geometries of (a) cylindrical (b) spherical and (c) jointed-arm manipulators.

Hence the coordinates of the end-effector are

$$\left.\begin{array}{l} x = H \cos \phi \\ y = H \sin \phi \\ z = z \end{array}\right\} \tag{2.7}$$

In the case of the spherical robot θ and ϕ are the rotational inputs and L the linear one. Hence, from Figure 2.14(b),

$$\left.\begin{array}{l} x = L \cos \theta \cos \phi \\ y = L \cos \theta \sin \phi \\ z = L \sin \theta \end{array}\right\} \tag{2.8}$$

In the jointed-arm robot all the inputs ϕ, θ_1 and θ_2 are rotary. Hence

$$\left.\begin{array}{l} x = \{L_1 \cos \theta_1 + L_2 \cos (\theta_1 + \theta_2)\} \cos \phi \\ y = \{L_1 \cos \theta_1 + L_2 \cos (\theta_1 + \theta_2)\} \sin \phi \\ z = L_1 \sin \theta_1 + L_2 \sin (\theta_1 + \theta_2) \end{array}\right\} \tag{2.9}$$

The increasing complexity of these equations confirms our earlier statement that control problems become more difficult as we move from the cartesian to the jointed arm configuration. Let us pursue this.

The inverse problem

Given the machine coordinates ϕ, θ, z, L, H, eqns (2.7) to (2.9) can be used to determine the 'world' coordinates x, y, z of the end-effector. In many cases however, we need to solve the inverse problem: given the world coordinates of an object or workpiece we wish to know the machine coordinates required to bring the end-effector to that particular point in space. This is straightforward in the case of the cartesian robot, since world coordinates are the same as machine coordinates. It becomes progressively more difficult as we move towards the jointed-arm robot.

For the cylindrical robot the required machine coordinates are

$$\left.\begin{array}{l} H = \sqrt{x^2 + y^2} \\ z = z \\ \phi = \tan^{-1} y/x \end{array}\right\} \tag{2.10}$$

For the spherical robot they are

$$\left.\begin{array}{l} L = \sqrt{x^2 + y^2 + z^2} \\ \theta = \tan^{-1} (z/\sqrt{x^2 + y^2}) \\ \phi = \tan^{-1} (x/y) \end{array}\right\} \tag{2.11}$$

The inverse transform is more difficult in the case of the jointed-arm robot, and we shall have to seek the help of Figure 2.15(a) which shows a jointed-arm robot viewed in the plane of the arm (see Figure 2.14(c)).

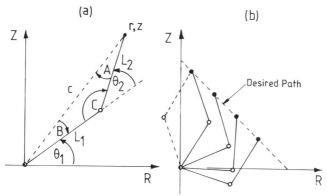

Figure 2.15 Solving the inverse problem for a jointed-arm robot: (a) the geometry in the plane of the arm (b) machine coordinates for a desired path.

Various trigonometrical relationships can be identified:

$$\left.\begin{aligned}
\phi &= \tan^{-1}(y/x) \\
r &= \sqrt{x^2 + y^2} \\
c &= \sqrt{r^2 + z^2} \\
A &= \cos^{-1}[(L_2^2 + c^2 - L_1^2)/2L_2c] \\
B &= \sin^{-1}[(L_2/L_1)\sin A] \\
\theta_1 &= \tan^{-1}(z/r) - B \\
\theta_2 &= A + B
\end{aligned}\right\} \qquad (2.12)$$

Hence for a given set of world coordinates these formulae can be applied sequentially to determine the machine coordinates ϕ, θ_1 and θ_2.

The results of such a set of calculations are shown in Figure 2.15(b). The desired path of the end-effector is assumed to be a straight line in the z, r-plane. The machine coordinates, and the resultant arm configurations, are shown for four points along this path. For each configuration there is an alternative and one of these is shown in dotted form.

The systems presented in Figure 2.14 and 2.15 provide control of three degrees of freedom, allowing the end-effector to be positioned in three-dimensional space. But a versatile robot requires six degrees of freedom so that orientation, as well as position, can be controlled. Let us see how the complexity of the inverse problem increases as the number of degrees of freedom is increased from 3 to 4.

Figure 2.16(a) shows an articulated arm with three elements, L_1, L_2 and L_3. It represents a jointed arm to which has been added a wrist with a single degree of freedom, i.e. the wrist can only rotate in the plane of the arm. The end-effector, or fingers, are at distance L_3 from the wrist. Now this additional degree of freedom allows the end-effector to be positioned at a point r, z in space *and* with a specified orientation ξ in the r, z-plane.

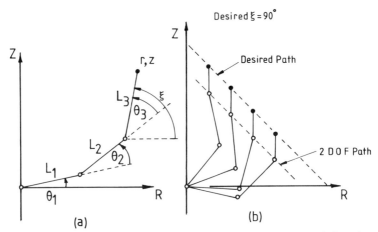

Figure 2.16   Solving the inverse problem for three degrees of freedom: (a) the geometry, showing the pitch angle ξ (b) machine coordinates for a desired path and orientation.

Knowing the orientation ξ and the length L_3 it is possible to determine the required position of the wrist r_w, z_w. From Figure 2.15(a),

$$\left.\begin{array}{l} z_w = z - L_3 \sin \xi \\ r_w = r - L_3 \cos \xi \end{array}\right\} \tag{2.13}$$

Knowing these coordinates, eqn (2.12) can be used to determine ϕ, θ_1 and θ_2. Figure 2.17(b) shows the results of such calculations for the case $L_1 = L_2 = 2L_3$ and $\xi = 90°$. Again, as in Figure 2.16(b) you will note that there are two possible arrangements for the upper and lower arm for each wrist position.

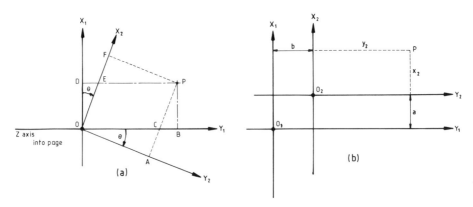

Figure 2.17   Transformations: (a) rotation of a frame (b) translation of a frame.

2.5 Matrix methods

Complexity increases rapidly with the number of degrees of freedom. For six degrees of freedom even the forward transform from machine coordinates to world coordinates, which is considered to be the easy one, can be very complex. The inverse problem can involve a lot of calculation and requires the analyst to exercise intuition in the resolution of many of its sub-problems, such as deciding which of several alternative geometrical con-figurations to choose. (See Figure 2.15(b).)

A knowledge of matrix methods is desirable if the kinematics of robots, particularly the more complex ones, are to be thoroughly understood. When the manipulator geometry is simple we have shown that a trigonometric solution may be available, but in many cases the trigonometric solution is found to be too difficult to apply to the arrangement with the full six degrees of freedom.

When using matrix methods (Paul, 1982), the initial step is to fix coordinate frames to each link of the manipulator, so that each frame rotates and translates with its link. If the cartesian coordinates of a point on a link are known with respect to its particular frame, say those of the end-effector, then transformation matrices can be used to transform the coordinates relative to any other frame. By this means, a series of transformations, depending on the number of links, can ultimately give the 'world' coordinates of any point in the chain.

Transformations

Let us examine a few examples of frame transformations and their relevant transformation matrices: firstly, a rotation of an orthonormal XYZ frame about its Z-axis. Figure 2.17(a) shows the frame in its original position $(XYZ)_1$ and in position $(XYZ)_2$ after a positive rotation θ around the z-axis. The coordinates of P in the two frames are related by

$$\left. \begin{array}{l} x_1 = x_2 \cos \theta - y_2 \sin \theta \\ y_1 = x_2 \sin \theta + y_2 \cos \theta \\ z_1 = z_2 \end{array} \right\} \tag{2.14}$$

These relationships can be expressed in a much more convenient form by using a homogeneous transformation matrix. (Denavit and Hartenberg, 1955; Paul *et al.*, 1981).

Homogeneous coordinates take the form (x, y, z, w) where x, y, z are the standard cartesian coordinates and w is a scaling factor, taken to be unity throughout this chapter. Using this terminology, eqns (2.14) can be written as

$$\begin{bmatrix} x \\ y \\ z \\ 1 \end{bmatrix}_1 = \begin{bmatrix} C\theta & -S\theta & 0 & 0 \\ S\theta & C\theta & 0 & 0 \\ 0 & 0 & 1 & 0 \\ 0 & 0 & 0 & 1 \end{bmatrix} \begin{bmatrix} x \\ y \\ z \\ 1 \end{bmatrix}_2 \qquad (2.15)$$

where $C\theta = \cos\theta$ and $S\theta = \sin\theta$.

Similarly, for a positive rotation α about the X-axis, it is easy to show (Paul, 1982) that

$$\begin{bmatrix} x \\ y \\ z \\ 1 \end{bmatrix}_1 = \begin{bmatrix} 1 & 0 & 0 & 0 \\ 0 & C\alpha & -S\alpha & 0 \\ 0 & S\alpha & C\alpha & 0 \\ 0 & 0 & 0 & 1 \end{bmatrix} \begin{bmatrix} x \\ y \\ z \\ 1 \end{bmatrix}_2 \qquad (2.16)$$

and for a positive rotation ϕ about the Y-axis

$$\begin{bmatrix} x \\ y \\ z \\ 1 \end{bmatrix}_1 = \begin{bmatrix} C\phi & 0 & S\phi & 0 \\ 0 & 1 & 0 & 0 \\ -S\phi & 0 & C\phi & 0 \\ 0 & 0 & 0 & 1 \end{bmatrix} \begin{bmatrix} x \\ y \\ z \\ 1 \end{bmatrix}_2 \qquad (2.17)$$

In addition to rotations about an axis we will also have to deal with the bodily translation of one frame with respect to another. If the frame $(XYZ)_1$ is moved bodily through distances a in a positive x-direction, b in a positive y-direction and c in a positive z-direction, then we can write

$$\left. \begin{array}{l} x_1 = x_2 + a \\ y_1 = y_2 + b \\ z_1 = z_2 + c \end{array} \right\} \qquad (2.18)$$

Figure 2.17(b) illustrates this for the two-dimensional case $(z = 0)$.

The homogeneous transform representation is then

$$\begin{bmatrix} x \\ y \\ z \\ 1 \end{bmatrix}_1 = \begin{bmatrix} 1 & 0 & 0 & a \\ 0 & 1 & 0 & b \\ 0 & 0 & 1 & c \\ 0 & 0 & 0 & 1 \end{bmatrix} \cdot \begin{bmatrix} x \\ y \\ z \\ 1 \end{bmatrix}_2 \qquad (2.19)$$

We are now in a position to recognize another feature of a homogeneous transform matrix T. It can be written in the general form

$$T = \left| \begin{array}{ccc|c} & R & & P \\ \hline 0 & 0 & 0 & 1 \end{array} \right| \qquad (2.20)$$

where R denotes the 3×3 rotational component of frame 2 with respect to frame 1 and P is a 3-element column vector pointing from the origin of frame 1 to the origin of frame 2.

Combined transformations

It may not be very obvious what this has to do with robot geometry. The reader's patience will soon be rewarded, but there is still one aspect of the matrix method to be understood; that of combined transformations. Let us assume for example that, relative to its initial position $(XYZ)_1$, the frame has been rotated in a positive manner about Z, translated by (L, O, D) and then rotated in a positive sense about X. What is the resultant overall transformation between $(XYZ)_1$ and $(XYZ)_2$? The answer can be found by taking the individual transformations in the correct sequence: here

$$\begin{bmatrix} x \\ y \\ z \\ 1 \end{bmatrix}_1 = (\text{rot } Z, \ \theta)(\text{tran } (L, O, D))(\text{rot } X, \ \alpha) \begin{bmatrix} x \\ y \\ z \\ 1 \end{bmatrix}_2 \tag{2.21}$$

or

$$\begin{bmatrix} x \\ y \\ z \\ 1 \end{bmatrix}_1 = \begin{bmatrix} C\theta & -S\theta & 0 & 0 \\ S\theta & C\theta & 0 & 0 \\ 0 & 0 & 1 & 0 \\ 0 & 0 & 0 & 1 \end{bmatrix} \begin{bmatrix} 1 & 0 & 0 & L \\ 0 & 1 & 0 & 0 \\ 0 & 0 & 1 & D \\ 0 & 0 & 0 & 1 \end{bmatrix} \begin{bmatrix} 1 & 0 & 0 & 0 \\ 0 & C\alpha & -S\alpha & 0 \\ 0 & S\alpha & C\alpha & 0 \\ 0 & 0 & 0 & 1 \end{bmatrix} \begin{bmatrix} x \\ y \\ z \\ 1 \end{bmatrix}_2$$

Multiplying the matrices gives

$$\begin{bmatrix} x \\ y \\ z \\ 1 \end{bmatrix}_1 = \begin{bmatrix} C\theta & -S\theta C\alpha & S\theta S\alpha & LC\theta \\ S\theta & C\theta C\alpha & -C\theta S\alpha & LS\theta \\ 0 & S\alpha & C\alpha & D \\ 0 & 0 & 0 & 1 \end{bmatrix} \begin{bmatrix} x \\ y \\ z \\ 1 \end{bmatrix}_2 \tag{2.22}$$

or

$$\left. \begin{aligned} x_1 &= x_2 C\theta - y_2 S\theta C\alpha + z_2 S\theta S\alpha + LC\theta \\ y_1 &= x_2 S\theta + y_2 C\theta C\alpha - z_2 C\theta S\alpha + LS\theta \\ z_1 &= y_2 S\alpha + Z_2 C\alpha + D \end{aligned} \right\} \tag{2.22}$$

Hence knowing the coordinates of a point relative to the frame $(XYZ)_2$, its coordinates relative to the original frame $(XYZ)_1$ can be determined. It should be emphasized that the order of events is important. If the order is changed the overall transformation will be different, since matrix multiplication is a non-commutative process.

The Denavit–Hartenberg matrix

The transformation matrix of eqn (2.22), illustrated the result of combined rotations and translations. It has a particular significance, being known as the Denavit–Hartenberg matrix. Let us see why it is so important in robot kinematics.

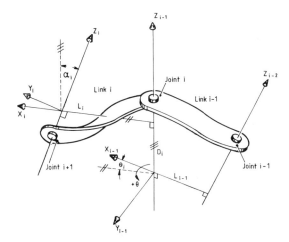

Figure 2.18   A spatial linkage.

Denavit and Hartenberg (1955) showed that any two neighbouring frames can be brought into coincidence by a prescribed sequence of at most two rotations and two translations. In order to understand this, it is necessary to study the space linkage shown in Figure 2.18. An orthonormal coordinate frame $(XYZ)_i$ is assigned to link i, where the Z_i-axis lies along the axis of rotation of joint $i + 1$, and the X_i-axis is normal to the Z_{i-1}-axis pointing away from it, while the Y_i-axis is at right angles to the other two. Link i is defined by two particular parameters: L_i, the common normal distance between the Z_{i-1}-axis and the Z_i-axis, and α_i, the twist angle measured between the Z_{i-1}-and the Z_i- axes in a plane perpendicular to L_i. Joint i has two important parameters: the distance D_i measured between the X_{i-1}- and X- axes and a revolute joint variable θ_i, which is the joint angle between the normals measured in a plane normal to the joint axis.

Having fixed frames to all links according to this scheme, the relationship between successive frames can be determined as follows. Working forward from frame $(XYZ)_{i-1}$ to $(XYZ)_i$, the rotations and translations needed to make the frames coincide can be identified. It is necessary:

(1) to rotate frame $i - 1$ through positive θ around the Z-axis in order to align the X-axes
(2) to translate in the positive sense along Z_{i-1} by D_i
(3) to translate in the positive sense along X_i by a distance L_i
(4) to rotate through positive α around the X_i axis in order to align the Z-axes.

The reader may have recognized this series of transformations to be the same as that dealt with earlier. The overall transformation is thus

$$\begin{bmatrix} x \\ y \\ z \\ 1 \end{bmatrix}_{i-1} = \begin{bmatrix} C\theta_1 & -C\alpha_1 S\theta_1 & S\alpha_1 S\theta_1 & L_1 C\theta_1 \\ S\theta_1 & C\alpha_1 C\theta_1 & -S\alpha_1 C\theta_1 & L_1 S\theta_1 \\ 0 & S\alpha_1 & C\alpha_1 & D_1 \\ 0 & 0 & 0 & 1 \end{bmatrix} \begin{bmatrix} x \\ y \\ z \\ 1 \end{bmatrix}_i \qquad (2.24)$$

Matrix methods applied to a robot with $M = 3$

The 4×4 matrix is often referred to as the A matrix. Let us demonstrate its use by examining the three-link manipulator shown in Figure 2.19, representing a trunk carrying an anthropomorphic arm. The trunk rotates about a vertical axis. The coordinate frames are established first, choosing axes of rotation for the Z-axes. Link 1 rotates through θ_1 about Z_0; link 2 rotates through θ_2 about Z_1; link 3 rotates through θ_3 about Z_2. Frame 0 is fixed to earth, frame 1 to link 1, frame 2 to link 2 and frame 3 to link 3.

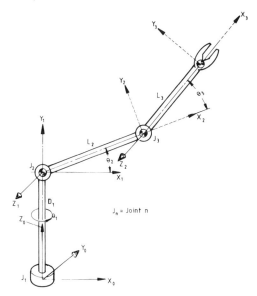

Figure 2.19 A manipulator ($M = 3$) showing coordinate frames.

Inspection of Figure 2.19 allows the following conclusions to be drawn: Firstly, in moving frame 0 into coincidence with frame 1:

(1) it has be rotated by θ_1 in the positive sense around axis Z_0
(2) it has to be translated along Z_0 by D_1
(3) there is no need for translation along $X_1 (L_1 = 0)$
(4) it has to be rotated by 90° in the positive sense around axis $X_1 (\alpha = 90°)$

Following this procedure for all three links allows the Table 2.1 to be constructed.

Table 2.1

Link	α	L	D	θ
1	90°	0	D_1	θ_1
2	0	L_2	0	θ_2
3	0	L_3	0	θ_3

The A matrices can now be constructed as follows:

$$_0A_1 = \begin{bmatrix} C1 & 0 & S1 & 0 \\ S1 & 0 & -C1 & 0 \\ 0 & 1 & 0 & D_1 \\ 0 & 0 & 0 & 1 \end{bmatrix} \quad _1A_2 = \begin{bmatrix} C2 & -S2 & 0 & L_2C2 \\ S2 & C2 & 0 & L_2S2 \\ 0 & 0 & 1 & 0 \\ 0 & 0 & 0 & 1 \end{bmatrix}$$

$$_2A_3 = \begin{bmatrix} C3 & -S3 & 0 & L_3C3 \\ S3 & C3 & 0 & L_3S3 \\ 0 & 0 & 1 & 0 \\ 0 & 0 & 0 & 1 \end{bmatrix}$$

where $C1 = \cos \theta_1$, etc. and

$$\begin{bmatrix} x \\ y \\ z \\ 1 \end{bmatrix}_0 = (_0A_1)(_1A_2)(_2A_3) \begin{bmatrix} x \\ y \\ z \\ 1 \end{bmatrix}_3$$

In general the overall transformation matrix T is defined by

$$_nT_m = (_nA_{n+1})(_{n+1}A_{n+2}) \cdots (_{m-1}A_m) \tag{2.25}$$

In this case

$$_0T_3 = (_0A_1)(_1A_2)(_2A_3)$$

and matrix multiplication gives

$$_0T_3 = \begin{bmatrix} C1C23 & -C1S23 & S1 & C1(L_2C2+L_3C23) \\ S1C23 & -S1S23 & -C1 & S1(L_2C2+L_3C23) \\ S23 & C23 & 0 & (D_1+L_2S2+L_3S23) \\ 0 & 0 & 0 & 1 \end{bmatrix} \tag{2.26}$$

where $C23 = \cos (\theta_2 + \theta_3)$, etc.

Hence, for example, the $(XYZ)_0$ coordinates of the tip of link 3 are

$$\left. \begin{aligned} x_0 &= C1(L_2C2+L_3C23) \\ y_0 &= S1(L_2C2+L_3C23) \\ z_0 &= D_1+L_2S2+L_3S23 \end{aligned} \right\} \tag{2.27}$$

This result agrees with that derived earlier in eqns (2.9) and illustrated in Figure 2.14.

This is an appropriate point to refer to prismatic joints. The Denavit–Hartenberg matrix developed earlier, and the example above, referred only to revolute joints, but this should not lead the reader to the conclusion that the method cannot handle prismatic joints. Constraints of space do not allow a proper discussion of the topic, and the interested reader is directed to the relevant literature (Paul, 1982; Paul *et al.*, 1981). However, it is worth noting two important differences between the matrix treatment of revolute and prismatic joints. For the latter the distance D is the joint variable, and the length L has no meaning and is hence set to 0.

Matrix methods applied to a robot with $M = 6$

We will use a commercial robot with six degrees of freedom for our last example of the use of A matrices. The Unimate Puma is sketched in Figure 2.20. Its jointed arm has three degrees of freedom and the hand (Figure 2.20(b)), has a further three degrees of freedom relative to the wrist. (We shall examine other wrist configurations later.) For each degree of freedom

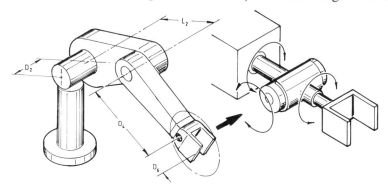

Figure 2.20 The Unimate Puma, showing wrist detail.

we have to establish coordinate frames, attached to appropriate links. The method of assigning these frames is shown in Figure 2.21 where, as before, the Z-axes are placed along the axes of rotation. Link numbers are shown in the rectangles. The necessary rotations and translations are summarized in Table 2.2

Table 2.2

Link	α	L	D	θ
1	−90	0	0	θ_1
2	0	L_2	D_2	θ_2
3	90	0	0	θ_3
4	−90	0	D_4	θ_4
5	90	0	0	θ_5
6	0	0	D_6	θ_6

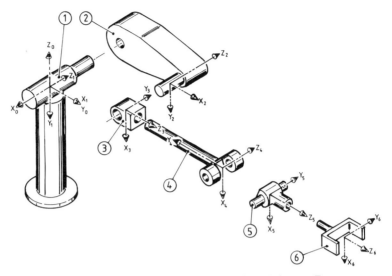

Figure 2.21 Coordinate frames for the Unimate Puma.

The positioning and orientating of frames is the most difficult part of this technique. One has to ensure that frame $i-1$ can be made coincident with frame i by rotation about z_{i-1}, translation along z_{i-1}, translation along x_i, and finally, rotation about x_i. Note that, for convenience, the origins of frame 0 and 1 are chosen to be coincident.

With the help of Table 2.2 and eqn (2.24) the A matrices can now be constructed:

$$_0A_1 = \begin{bmatrix} C1 & 0 & -S1 & 0 \\ S1 & 0 & C1 & 0 \\ 0 & -1 & 0 & 0 \\ 0 & 0 & 0 & 1 \end{bmatrix} \qquad _1A_2 = \begin{bmatrix} C2 & -S2 & 0 & L_2C2 \\ S2 & C2 & 0 & L_2S2 \\ 0 & 0 & 1 & D_2 \\ 0 & 0 & 0 & 1 \end{bmatrix}$$

$$_2A_3 = \begin{bmatrix} C3 & 0 & S3 & 0 \\ S3 & 0 & -C3 & 0 \\ 0 & 1 & 0 & 0 \\ 0 & 0 & 0 & 1 \end{bmatrix} \qquad _3A_4 = \begin{bmatrix} C4 & 0 & -S4 & 0 \\ S4 & 0 & C4 & 0 \\ 0 & -1 & 0 & D_4 \\ 0 & 0 & 0 & 1 \end{bmatrix}$$

$$_4A_5 = \begin{bmatrix} C5 & 0 & S5 & 0 \\ S5 & 0 & -C5 & 0 \\ 0 & 1 & 0 & 0 \\ 0 & 0 & 0 & 1 \end{bmatrix} \qquad _5A_6 = \begin{bmatrix} C6 & -S6 & 0 & 0 \\ S6 & C6 & 0 & 0 \\ 0 & 0 & 1 & D_6 \\ 0 & 0 & 0 & 1 \end{bmatrix}$$

The overall transformation between frame 6 and frame 0 is given by

$$_0T_6 = (_0A_1)(_1A_2)(_2A_3)(_3A_4)(_4A_5)(_5A_6)$$

$$= \begin{bmatrix} n_x & s_x & a_x & p_x \\ n_y & s_y & a_y & p_y \\ n_z & s_z & a_z & p_z \\ 0 & 0 & 0 & 1 \end{bmatrix}$$

(2.28)

where (see Lee and Ziegler (1983))

$n_x = C1[C23(C4C5C6 - S4S6) - S23S5S6] - S1[S4C5C6 + C4S6]$

$n_y = S1[C23(C4C5C6 - S4S6) - S23S5C6] + C1[S5C5C6 + C4S6]$

$n_z = -S23[C4C5C6 - S4S6] - C23S5C6$

$s_x = C1[-C23(C4C5S6 + S4C6) + S23S5S6] - S1[C4C6 - S4C5S6]$

$s_y = SI[-C23(C4C5S6 + S4C6) + S23S5S6] + C1[C4C6 - S4C5S6]$

$s_z = S23(C4C5S6 + S4C6) + C23S5S6$

$a_x = C1(C23C4S5 + S23C5) - S1S4S5$

$a_Y = S1(C23C4S5 + S23C5) + C1S4S5$

$a_z = C23C5 - S23C4S5$

$p_X = C1[D_6(C23C4S5 + S23C5) + S23D_4 + L_2C2] - S1(D_6S4S5 + D_2)$

$p_y = S1[D_6(C23C4S5 + S23C5) + S23D_4 + L_2C2] + C1(D_6S4S5 + D_2)$

$p_z = D_6(C23C5 - S23C4S5) + C23D_4 - L_2S2$

The inverse problem with $M = 6$

In the case we have just dealt with it was clear that even the forward transformation from machine coordinates to world coordinates requires a lot of computation. The inverse problem is even more daunting. Several investigations (Paul *et al.*, 1981; Lee and Ziegler (1983)) have used methods based on trigonometry, starting at the end-effector and work backwards. These methods are an extension of the techniques we discussed earlier and illustrated in Figure 2.19. Others (Milenkovic, 1979) have developed iterative techniques which require the robot to be conceptually divided into two linkages. One of these, the major linkage, consists of the first three joints and the related links: it produces gross motion of the end-effector. We have seen several examples of such major linkages in Figure 2.10. The other linkage, the wrist, orientates the end-effector. The iterative procedure alternately computes and adjusts the machine coordinates of the major linkage to correct position, whilst holding the wrist fixed, and adjusts the rotations of the wrist to correct the orientation, whilst holding the spatial coordinates of the wrist fixed. This process has been found to converge quickly.

The complexity of the inverse problem is another reason for moving away from serial-operated systems to parallel-operated systems like those shown in Figure 2.10. Equation (2.6) showed how one of the machine coordinates could be determined knowing the world coordinates. Similar relatively simple, although long-winded, expressions can be formulated for systems with six degrees of freedom (Stewart, 1965).

Differentials

The difficulty of solution of the inverse problem can be eased by using small perturbations or differential movements to linearize the equations. Consider, for example, the jointed-arm robot of Figure 2.19. We saw that coordinates in frame 0 are related to those in frame 3 by the transformation

$$
\begin{bmatrix} x \\ y \\ z \\ 1 \end{bmatrix}_0 = T \begin{bmatrix} x \\ y \\ z \\ 1 \end{bmatrix}_3
$$

where T was given in eqn (2.26).

The question we now ask is, given the coordinates of a point in frame 3, what perturbations will occur in its coordinates in frame 0 if there are perturbations in the machine coordinates θ_1, θ_2 and θ_3? In order to simplify matters let us consider the point $x_3 = y_3 = z_3 = 0$. We then have

$$
\left.
\begin{array}{l}
x_0 = C1(L_2 C2 + L_3 C23) \\
y_0 = S1(L_2 C2 + L_3 C23) \\
z_0 = D_1 + L_2 S2 + L_3 S23
\end{array}
\right\}
\tag{2.29}
$$

If small perturbations in θ_1, θ_2 and θ_3 occur then the resulting perturbations in x_0, y_0 and z_0 will be

$$
\begin{bmatrix} \Delta x_0 \\ \Delta y_0 \\ \Delta z_0 \end{bmatrix} =
\begin{bmatrix}
\dfrac{\partial x_0}{\partial \theta_1} & \dfrac{\partial x_0}{\partial \theta_2} & \dfrac{\partial x_0}{\partial \theta_3} \\[2ex]
\dfrac{\partial y_0}{\partial \theta_1} & \dfrac{\partial y_0}{\partial \theta_1} & \dfrac{\partial y_0}{\partial \theta_3} \\[2ex]
\dfrac{\partial z_0}{\partial \theta_1} & \dfrac{\partial z_0}{\partial \theta_1} & \dfrac{\partial z_0}{\partial \theta_3}
\end{bmatrix}
\begin{bmatrix} \Delta \theta_1 \\ \Delta \theta_2 \\ \Delta \theta_3 \end{bmatrix}
\tag{2.30}
$$

The 3×3 matrix is sometimes referred to as the Jacobian, J.

The derivatives $\partial x_0 / \partial \theta_1$, etc., are determined by differentiation of eqns

(2.29). For example,

$$\frac{\partial x_0}{\partial \theta_1} = -(L_2 C2 + L_3 C23)S1$$

$$\frac{\partial x_0}{\partial \theta_2} = -(L_2 S2 + L_3 S23)C1$$

$$\frac{\partial x_0}{\partial \theta_3} = (L_2 C2 - L_3 S23)C1$$

Hence it is possible to evaluate J given initial values of θ_1, θ_2 and θ_3. For example, with $L_2 = L_3 = 1$; $\theta_1 = 30°$; $\theta_2 = 60°$ and $\theta_3 = 30°$ the Jacobian becomes

$$J = \begin{bmatrix} -0.250 & -1.616 & -0.866 \\ 0.433 & -0.933 & -0.500 \\ 0 & 0.5 & 0 \end{bmatrix}$$

Hence for known perturbations in θ_1, θ_2 and θ_3 the resultant perturbations in x, y, and z can be calculated, i.e. the forward problem. For example if $\Delta\theta_1 = \Delta\theta_2 = \Delta\theta_3 = 0.05$ rad then

	Approximation	Exact
$\Delta x =$	-0.1316	-0.1337
$\Delta y =$	-0.0500	-0.0566
$\Delta z =$	0.0250	0.0189

The exact values were calculated using eqns (2.29). The accuracy of the approximation will improve as $\Delta\theta$ decreases.

However, we are here concerned with the more difficult inverse problem—knowing Δx, Δy, Δz, what are $\Delta\theta_1$, $\Delta\theta_2$ and $\Delta\theta_3$? To solve this problem, eqn (2.30) has to be rewritten as

$$\begin{bmatrix} \Delta\theta_1 \\ \Delta\theta_2 \\ \Delta\theta_3 \end{bmatrix} = J^{-1} \begin{bmatrix} \Delta x \\ \Delta y \\ \Delta z \end{bmatrix} \qquad (2.31)$$

where J^{-1} is the inverse of J. The inverse of a square matrix J is given by

$$J^{-1} = \frac{\text{adj}(J)}{(J)}$$

where adj (J) is the adjoint of J and (J) is the determinant of J. In this case

$$J^{-1} = -4 \begin{bmatrix} 0.250 & -0.433 & 0 \\ 0 & 0 & -0.500 \\ 0.2165 & 0.125 & 0.933 \end{bmatrix}$$

and taking perturbations $\Delta x = \Delta y = \Delta z = 0.025$ gives

	Approximation	Exact
$\Delta\theta_1 =$	1.049°	0.981°
$\Delta\theta_2 =$	2.865°	3.496°
$\Delta\theta_3 =$	−7.303°	−8.130°

The exact values were derived from eqns (2.12). Once again the errors are seen to be significant, but they can be reduced by making the perturbations Δx, Δy, and Δz smaller. In the worked example the perturbations were relatively large, i.e. 2.5 cm for a link length of 1 m.

Thus, although this method reduces the complexity of the inverse problem it introduces inaccuracies which can be reduced only by using very small perturbations. If such a technique were used to guide an end-effector along a continuous path, the path would have to be divided into many small sections, and movement along each would then be determined by the time required for computation.

There are applications where extreme accuracy is not important, such as prosthetic arms or tele-operated devices (see Chapter 10), and in such cases this technique is useful. There are other instances where, by the nature of the problem, the perturbations are very small and the technique is satisfactory. Such an instance is the allowance for movements caused by the compliance of a robot's structure (Paul, 1982). Finally, the method is directly applicable when speed control is used, i.e. the velocity of the end-effector in space is proportional to the rate of change of the machine coordinates. We shall see more of this in Chapter 5.

2.6   Recapitulation

This chapter has attempted to demonstrate how the number of degrees of freedom of a robot's mechanism can be evaluated. Mechanisms for serial operation and for parallel operation have been identified; although the vast majority of existing robots use serial operation, parallel operation has many advantages to offer.

Trigonometry and matrix methods have been used to establish relationships between the world coordinates of the end-effector and the machine coordinates of the robot. Both forward and inverse problems have been discussed; the latter is particularly difficult when mobility is high. The method of differentials is helpful in the solution of the inverse problem.

Chapter 3

Wrists, hands, legs and feet

3.1 Introduction

In anthropomorphic terms the last chapter concentrated on the robot's trunk and arm. Now we examine the other essential parts of its anatomy. A wrist is needed to orientate the hand or end-effector in space, and the hand itself must be capable of grasping a tool or a workpiece. If mobility is desirable, the robot must be equipped with wheels, tracks or legs.

3.2 Wrists

Whilst the first three links of a robot are used for gross position control of an end-effector, the main function of the wrist assembly is angular orientation. Some of the simpler robots do not have wrists: they are used in applications where tooling sets the workpiece at a fixed orientation relative to the robot. However, most tasks will require at least one angular orientation and many will require the maximum of three. Figure 3.1 shows two peg-in-hole assembly tasks (Potkonjac *et al.*, 1983)—a cylinder into a circular hole and a cuboid into a square hole. The former requires the correct adjustment of two independent angular rotations, ε and ξ. In the latter case the cuboid, lacking the axial symmetry of the cylinder, requires three orientations, ε, ξ and ρ. Note that these angles are world coordinates.

Each independent orientation of the end-effector requires a corresponding degree of freedom in the wrist assembly. The three most commonly used terms for end-effector orientation are pitch, yaw and roll. *Pitch* is rotation of the end-effector about a horizontal axis at the end of the

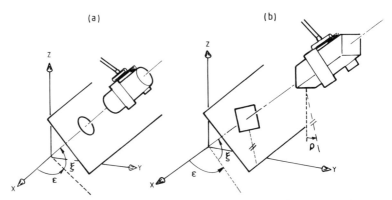

Figure 3.1 Required orientational freedoms: (a) yaw ε and pitch ξ are sufficient when object has axial symmetry (b) roll ρ is also required when there is no axial symmetry.

robot arm and perpendicular to its axis. It controls the angle ξ in Figure 3.1(b). It gives the end effector an up-and-down rotary motion. *Yaw* is rotation about a vertical axis perpendicular to the pitch axis. It gives the end-effector a side-to-side rotary motion and controls the angle ε in Figure 3.1(b). Finally, *roll* is rotation about the longitudinal axis of the wrist. It controls the angle ρ.

These motions are further illustrated in Figure 3.2. The first sketch (a) shows a side elevation of a wrist with three degrees of freedom, using three drive units, each with its axis perpendicular to those of the other two and each devoted to setting a particular orientation, pitch *P*, yaw *Y* and roll *R*.

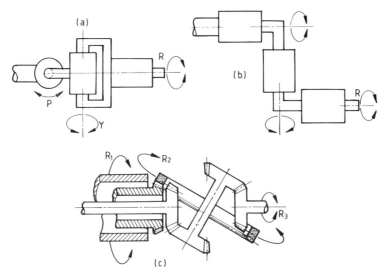

Figure 3.2 Wrists: (a) all axes mutually perpendicular (b) all elements capable of continuous rotation (c) a design that allows remote drive.

(*P, Y* and *R* are machine coordinates.) Note that the end-effector is attached to the roll axis: this allows rotary tools such as drills to be used. This design is not particularly compact and is difficult to drive remotely, usually requiring a direct drive with motors placed as shown: thus the arm has to carry these motors as an additional burden. However, large angular movements are possible and medium to heavy loads can be handled.

A different arrangement is shown in Figure 3.2(b). Here combined rotations about the first two axes yield the pitch and yaw orientation (Stackhouse, 1979; Colson and Perreira, 1983). The third axis gives roll. In contrast to (a), all three axes can give continuous rotation and are easily driven by remote drives. This assembly can be quite compact and can handle small to medium loads.

Before moving on to Figure 3.2(c) the reader is asked to refer back to Figure 2.20 which shows yet another compact three-degrees-of-freedom wrist. Again, unlike (a), there are no axes specific to pitch and yaw, and combined rotations about the first two axes provide these orientations. Large angular orientations are achievable. Since all three axes intersect, the assembly can be compact and can be driven remotely.

Returning to Figure 3.2(c) we see how a wrist can be driven remotely (Stackhouse, 1979). Concentric shafts are driven by remote motors. Motor shaft R_1 directly rotates the assembly containing the R_2- and R_3-axes, whilst the R_2-axis rotates the R_3-structure along a conical surface in space. Rotation about each axis is continuous and reversible. The remote drive, and the fact that all three axes intersect, allows for a very compact design.

3.3 Grippers

There are many ways of providing an end-effector with six degrees of freedom, but none of them serves a useful purpose until an actual payload is involved. This can take the form of a tool, such as a welding torch, a drill, a spraygun, etc. Alternatively the payload can be an object, such as a casting, a chocolate, a sheet of glass, which is being transferred from one work station to another.

We shall bypass the rather specialized tool holder and focus our attention on the gripper, whose main function is to grasp and release workpieces during transfer. Following Fan Yu Chen (1982) we shall classify grippers as mechanical, vacuum and magnetic, or universal.

Mechanical grippers

Mechanical grippers, according to Konstantinov (1975) can be further subdivided into linkage types, rack and pinion types, cam types, screw types, rope and pulley types, and miscellaneous types. Figure 3.3 shows examples of the first four types. The linkage type is illustrated in (a) to (d), the rack and pinion type in (e) and (f), the cam type in (g) and (h), and the screw

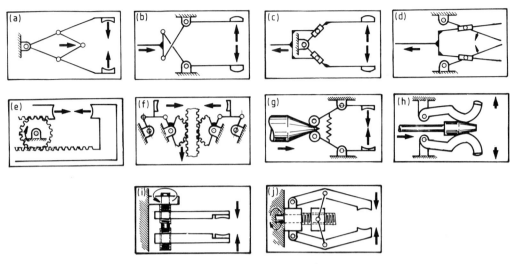

Figure 3.3 Mechanical grippers.

type in (i) and (j). The operation of these grippers is self explanatory, but some factors should be noted:

(a) and (b) use only revolute pairs
(c) and (d) use prismatic as well as revolute pairs
(b), (c) and (h) are used for internal gripping
(d) is known as the swing block mechanism
(g) uses a spring for release
(e), (i) and (j) require rotary inputs
(f) provides parallel motion of the gripper jaws
(g) and (h) can use a variety of cam profiles—constant velocity, harmonic, etc.

In all cases the input force can be supplied by means of pneumatic, hydraulic or electric motors or actuators. The gripper redirects and magnifies this force, the actual required force at the jaws being dependent on the weight of the component, the friction between the jaws and the component, the acceleration of the end-effector, the orientation of the component in the jaws, the relation between the direction of movement and this orientation.

Another interesting example of the use of mechanical linkages is the remote centre compliance (RCC) device (Whitney and Nevins, 1979). Although it is not a gripper, it is appropriate to refer to the RCC here since it helps the mechanical gripper to be more effective as an assembly tool. The RCC acts as a multi-axis 'float', allowing lateral and angular misalignments between parts to be accommodated. Figure 3.4 illustrates its principle of operation. The task is to insert a round peg into a round hole, allowing for the possibility of both linear and angular errors in the position of the peg with respect to the hole. The RCC is designed so that lateral error is not converted into angular error during engagement. Figure 3.4(a) shows how a

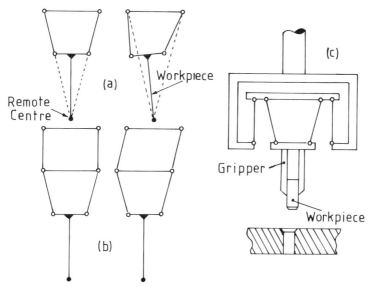

Figure 3.4 The remote centre compliance (RCC): (a) the linkage allows rotation about the remote centre (b) lateral offset is also accommodated (c) a practical arrangement.

peg is held so that it can rotate about its tip, that is, about the point where it engages a mating part. This allows the peg to realign itself in response to contact forces during insertion in the hole. The point of rotation is known as the remote centre, and it is obviously an unusual part of a coupler curve.

A unique feature of the RCC device is that it absorbs lateral error and angular error independently. Figure 3.4(b) shows how the addition of a parallelogram linkage allows lateral errors to be accommodated. In combination, the vertical links with their remote centre at infinity allow only lateral motion, whilst the lower linkage with its remote centre at the tip of the peg allows only angular rotation in response to moments applied at the tip. During assembly the lateral part is active during engagement whilst the angular part takes over during insertion. Figure 3.4(c) shows a compact, axisymmetric version of the RCC.

Fan Yu Chen's classification of mechanical grippers included a subsection for miscellaneous types. Two particular grippers of this type are worthy of mention: both use air pressure to change the shape of a flexible bag. The first, the expansion gripper (Lundstrom *et al.*, 1977), is often used when components have conveniently located holes or bores. The simplest of this type uses a polyurethane bladder which is pushed into a hole and then inflated (Neubauer, 1982). Holding forces of around 500 N can be generated. For delicate components such as glass or plastic, nitrile butadiene diaphragms operated at pressures around 2 bar can be used. External gripping can be achieved by means of expandable rubber rings, placed for example around the neck of a bottle and then pressurized.

The other gripper worthy of mention under the miscellaneous heading uses pneumatic fingers, each consisting of a hollow polyurethane body with serrations on one side (Neubauer, 1982). When compressed air is supplied these serrations expand, causing the finger to bend away from the serrations and over the part to be held. A two-fingered hand of this kind can hold a load of 60 N when operated at 5 bar.

Vacuum and magnetic operation

Vacuum cups (Lundstrom *et al.*, 1979; Neubauer, 1982) offer a convenient form of gripping where surfaces are flat and smooth, e.g. glass. Cups are of two basic types. One relies on the natural vacuum that is created when a cup is pressed on to a flat surface; maximum holding force for such cups is about 140 N. The other type is operated by a vacuum pump, and can be used on parts with rougher surfaces. Depending on the surface quality, forces of around 200 N are achievable. It is usual to employ several cups arranged in a pattern to suit the component.

Magnetic grippers (Chen, 1982; Neubauer, 1982) are often used when flat ferrous materials have to be handled. Permanent magnets or electro-magnets may be used. Permanent magnets do not require a power source and are therefore suitable for use in hazardous environments, but they have disadvantages: components have to be released mechanically, and ferrous rubbish can build up on the magnet and on the component. Both of these disadvantages can be overcome if electromagnets are used. Although requiring a power source they are well suited to remote control and have a moderately fast pick-up and release.

Universal grippers

Our final look at grippers reviews the most recent attempts to design one capable of handling a wide variety of shapes. The human hand is such a gripper, and we shall have more to say about it later.

Over recent years there has been a gradual evolution from the special-purpose gripper to the versatile gripper. An early step towards increased universality was the introduction of the double gripper (Reed, 1979), which allows both machined and unmachined components to be handled simultaneously during a machining cycle—two hands are better than one. Another significant advance towards the universal gripper is provided by those inflatable grippers (Chen, 1982) that can adapt their shape to a wide variety of part profiles. Such devices rely on the use of a flexible bag, filled with a loose material such as small spherical particles of glass or metal. This bag, placed around the component, is pressurized to make it conform to the component's shape. The bag is then evacuated and the loose material consolidates to hold the part firmly.

A more direct approach to universal gripping has been developed by

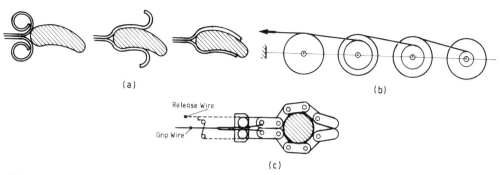

Figure 3.5 The soft gripper: (a) gripping action (b) pulleys provide uniform pressure (c) gripping a cylinder.

Lundstrom (1976) who uses a fast moulding technique to produce grippers for a variety of widely different shapes.

Another noteworthy evolutionary step in gripper design is the so-called soft gripper (Hirose and Umetani, 1977) (Figure 3.5). Inspired by observation of the movement of snakes, this device has a simple control system and can softly and gently conform to objects of any shape. During gripping it exerts a uniform pressure on the surface of the component. Figure 3.5(a) shows how an object is gradually enveloped by the soft gripper. As shown in (b) and (c), the gripper consists of a series of links, each with a torque applied by means of a grip wire. In order to obtain a uniform pressure distribution along the component, the torque applied to each link has to decrease as the tip is approached. (This is analogous to the variation of bending moment along a cantilever with uniform loading.) The required variation in torque is contrived by means of the pulley arrangement in (b).

Applications envisaged for the soft gripper include handling of glassware, fruit harvesting, egg handling, transporting hospital patients from bed to bed, and capturing wild animals.

Mechanical hands

The ultimate universal gripper is the human hand (Kato, 1977), but its complexity is so great that the prospect of seeing its technological equivalent is extremely remote. The bones of the human hand (Figure 3.6(a)) are subdivided into three segments: the carpus or wrist bones; the metacarpus or bones of the palm: and the phalanges or bones of the digits. There are eight bones of the carpus; five bones of the metacarpus, one for each digit; and fourteen phalanges, three for each finger and two for the thumb. In total these allow for about 22 major degrees of freedom.

For robotic application it has been concluded that three fingers would give adequate flexibility. Figure 3.6(b), (c) and (d) shows various designs of this kind. In (b) (Skinner, 1975) each finger, like the human hand, has three phalanges. All joints use simple R pairs. Two of the fingers, numbers 2 and

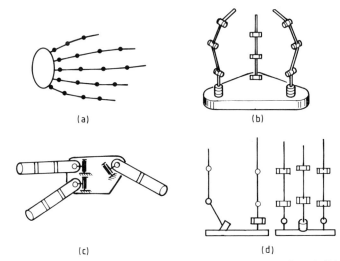

Figure 3.6 Mechanical hands: (a) the bones of the human hand (b) three fingers, each with three phalanges (c) two fingers and a thumb (d) another example of two fingers and a thumb.

3, can be rotated about vertical axes, thereby changing their bending directions. This allows the hand to perform a variety of prehensile tasks. For example, all three fingers can be made to curl around an object if their directions of bending are made parallel. If the bending directions are arranged symmetrically, however, the finger tips will meet in a pinching action when the fingers are bent.

Figure 3.6(c) (Okada and Tsuchiya, 1977) shows a mechanical hand that resembles the human hand more closely. There are two fingers and a thumb, each finger having three phalanges and the thumb two. These are intended to correspond to the index and middle fingers and the thumb of the human hand. Again, like the human hand, each digit can move as a whole from side to side, in addition to bending. The fingers are hollow; each joint is driven by cables which run through coil-like hoses so that the cables do not interfere with the complicated finger motions.

Figure 3.6(d) shows two views of another hand designed to mimic the motions of the thumb, index and middle fingers (Lian *et al.*, 1983). Once again, only simple R pairs are employed. This is a simpler design than (c), having two fewer degrees of freedom.

Finally we consider the mechanical hand shown in Figure 3.7 (Rovetta *et al.*, 1981). This is different from the others in that it has a palm which, when used in conjunction with the fingers, allows objects of varying geometries to be grasped firmly. In addition to the three conventional phalanges, each finger has a fourth link at its base. This element, called the inverse joint, reproduces hyperextension of the first phalanx of the human hand.

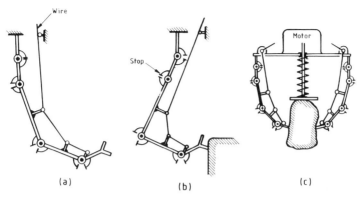

Figure 3.7 Mechanical hand with flexible palm: (a) approaching an object (b) contacting an object (c) gripping an object.

A preloaded spring and a mechanical stop are placed at each finger joint. The springs determine the relative movements of the phalanges as well as acting to restore the fingers to their resting position. Each phalanx is operated by a wire which changes both its position and its tension during finger movement. Figure 3.7(a) shows a finger approaching an object and (b) shows the change in configuration when the object is contacted.

Three fingers are used in conjunction with a spring-loaded palm. Figure 3.7(c) shows two fingers, but in practice three fingers are disposed symmetrically around the palm. When an object is placed amongst the fingers, its presence is sensed by an optical device which activates the stepping motors. The wires are placed in tension, the fingers touch the component and together with the palm they grasp it and lift it up. When the voltage from the position sensor on the palm exceeds a preset level the motors are stopped. Components up to 100 N in weight and of varying shapes have been grasped by this particular hand.

In concluding this section on mechanical hands, it is worth drawing attention to the anthropomorphic nature of the hands illustrated in Figures 3.6 and 3.7. We tend to try to copy our own mechanisms, which is not surprising, bearing in mind the versatility and effectiveness of the human machine. But the copies shown above are poor functional models of the human hand.

However, artificial hands are now becoming available with excellent cosmetic appearance and many degrees of freedom.

3.4 Mobile robots

Having dealt with the upper part of the robot's torso—its body, arms, wrists and hands we now concentrate on the lower half—the robot's legs and feet.

There are many instances where it is necessary to use a mobile robot. In flexible machining systems (FMS), in addition to developing flexible machines for assembly, machining, etc., it is equally important that there are flexible means of transportation amongst the various production processes. Fixed equipment such as rollers, belts and overhead conveyors is inflexible and can be expensive to change if machine layouts and transportation routes have to be varied. In recognition of this, automated guided vehicle systems (AGV) have become increasingly popular over the last 30 years. Such systems employ computer-controlled trucks or skips guided by embedded wires or reflective tapes. But greater flexibility can be achieved if, instead of restricting the vehicle's movement to a set of guideways, it can be taught the route by an operator, like any other robot.

In addition to FMS there are many other potential applications for mobile robots: the domestic robot to help with housework: the night-watchman robot for ensuring the security of buildings; the fire-fighting robot that can go right to the heart of the fire; the patient-care robot that can relieve the nurse of hard physical labour such as lifting, holding and carrying patients or handicapped children.

3.5 Methods of support: wheels and tracks

The wheel is well established as a means of support for vehicles. It is energy-efficient, particularly in railways where the wheel and the rail are hard and smooth. For road use, where surfaces are liable to be uneven, it is necessary to trade energy efficiency for comfort by using a compliant pneumatic wheel. In robot systems, where the route is preprogrammed and floors are even, it is usual to employ hard tyres. All-wheel steering provides maximum manoeuverability allowing the robot to move at right angles to a given path and to rotate about its centre of gravity. This is useful for moving in corridors and other restricted areas.

When programming a robot vehicle a transducer, typically an optical encoder, is used to count the number of rotations of the wheel. When there is no slip this of course gives a measure of the vehicle's position. However, if slip occurs the correspondence between wheel rotation and vehicle position is lost, and it is necessary to include some form of corrective action. This can be achieved (Fujiwara *et al.*, 1977) by using a gyroscope to detect the direction of motion of the vehicle. During teaching, the gyroscope measures the absolute direction and this information is stored in memory. When the robot is travelling automatically, measured information from the gyroscope is compared with the stored information and corrective action is taken to correct directional errors. Errors in distance covered can be corrected by providing the vehicle with optical sensors which 'lock on' to reflective plates at the arrival station.

Conventional wheeled robots are satisfactory for FMS applications where the terrain is flat and well-ordered, but useless in environments requiring the mounting of steps and staircases, or the circumvention of sizeable obstructions. In such cases it is necessary to use systems in which

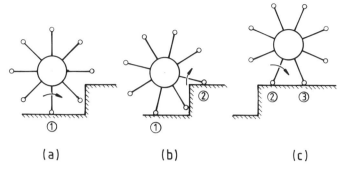

Figure 3.8 The rimless wheel mounting a step.

the point of support changes intermittently rather than continuously. Legged systems operate in this way, and we shall discuss these in more detail later. In the meantime let us examine the rimless wheel and the Venetian wheel.

Figure 3.8 shows how a rimless wheel with eight spokes surmounts a step. In (a) the wheel rotates about spoke 1 until spoke 2 contacts the top of the step as in (b). Rotation then takes place about spoke 2 until spoke 3 makes contact; and so on. Rimless wheels can climb steps with heights nearly equal to the radius of the wheel (Thring, 1983). They can give an uncomfortable ride, however; the centre of the wheel oscillates with an amplitude of $r(1 - \sin \pi/n)$ and a frequency of $n\omega$, where r is the radius of the wheel, n the number of spokes and ω the angular velocity of the wheel in rad/s.

The Venetian wheel is another device for mounting steps. It is so called because it is reputed to have its origins in Venice where porters use it to help them get their barrows up and down steps. It is available today from manufacturers of mechanical handling equipment, and in its commonest form it consists of three freely rotating wheels spaced at equal intervals around the circumference of a circle. Figure 3.9 shows a powered form in which the three wheels could be driven through idlers from a central gear wheel. On contacting the face of a step (Figure 3.9(a)), frictional forces cause wheel 1 to stop rotating and as a result the frame rotates clockwise

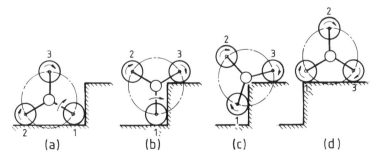

Figure 3.9 The Venetian wheel mounting a step.

about the axis of wheel 1. In Figure 3.9(b) wheel 3 makes contact with the top of the step. Figure 3.9(c) shows wheel 3 rolling along the top of the step, whilst wheel 1 rolls up the face of the step. A mobile robot, the TO-ROVER, with four such wheels, is used for inspection and repair of nuclear power plant and is described in Takano and Odawara (1981). But whereas the wheel in Figure 3.9 relies on friction for the rotation of the frame, the TO-ROVER uses a tachogenerator to detect when wheel 1 has stopped (Figure 3.9(a)) and then initiates a powered 120° clockwise rotation of the frame.

Tracked vehicles, such as tanks, are also capable of climbing steps, but they can impact the ground with considerable force when the body rotates after it moves across an obstacle. This impact is reduced to some extent in the fire-fighting robot (Takata and Kishimoto (1977)), in which, by attaching springs to the front and back axles in a horizontal direction, the caterpillar track can flex lengthwise and can therefore absorb the shock of impact. Thring (1983) shows several ways of eliminating these impacts. In one, each T-shaped element of the track performs like a leg. On approaching a step the leading edge of the track acts like a stepped wheel.

In addition to climbing over obstacles, the tracked vehicle can travel over soft ground that would be impassable to a conventional wheeled vehicle. It does this by distributing the load over a relatively large area. For soft ground it is necessary to keep the contact pressure below about $100 \, \text{kN/m}^2$. Contact pressures vary from around $41 \, \text{kN/m}^2$ for farm tractors to $170 \, \text{kN/m}^2$ for walking excavators. For comparison, a full-grown elephant exerts a pressure of around $80 \, \text{kN/m}^2$ on the ground (Hutchinson, 1967).

3.6 Walking machines

Walking machines offer several advantages over conventional wheeled vehicles. Whereas wheels have to be restricted to comparatively flat uniform surfaces, legged vehicles can traverse uneven terrain with a minimum of swinging and lurching. They can manoeuvre in confined spaces in buildings, and can climb stairs and they can haul heavy loads over soft ground. There is therefore a lot of interest nowadays in the development of walking machines for helping handicapped people and for industrial, agricultural and military use. Their mechanical complexity and their control problems have so far restricted developments to the laboratory, but nevertheless there is a lot to be gained and research continues.

Footfall sequences

In order to illustrate the complexity of walking machines, let us start by examining the possible variations in footfall sequence, including the simultaneous operation of two or more legs.

For a one-legged machine there is clearly only one sequence—the hop (snail and snake locomotion is not considered). For a biped, such as a human being, there are two possible sequences—the two-legged hop (the sack race) and the walk or run in which the legs operate sequentially. If the legs are numbered 1 and 2, we can describe these sequences as (1–2), and (1, 2) respectively, where a hyphen between numbers indicates simultaneous operation and a comma indicates sequential operation.

Using this notation we can list six possible sequences of leg operation for a tripod:

1–2–3 —the three-legged hop

$$\left.\begin{array}{l} 1, 2–3, \\ 2, 1–3, \\ 3, 1–2, \end{array}\right\} \text{two legs moving as one}$$

$$\left.\begin{array}{l} 1, 2, 3, \\ 2, 1, 3, \end{array}\right\} \text{one leg at a time: note that } (3, 1, 2,) \text{ and } (2, 3, 1,) \text{ are identical to } (1, 2, 3,) \text{ since over a series of cycles the order of leg operation is the same.}$$

For quadrupeds there are 26 possible sequences of leg operation. The first one (1–2–3–4,) is the four-legged hop in which the action is similar to that of a one-legged machine. The others are listed in Table 3.1. Bessonov and Umonov (1976) have used combinational methods to determine the number of leg sequences for any number of legs. Their results, for up to 10 legs, are given in Table 3.2.

Table 3.1

Operating effectively as a biped	*Operating effectively as a tripod*		*Quadruped*
1, 2–3–4,	1, 2, 3–4,	1, 3–4, 2,	1, 2, 3, 4,
2, 1–3–4,	1, 3, 2–4,	1, 2–4, 3,	1, 2, 4, 3,
3, 2–1–4,	1, 4, 2–3,	1, 2–3, 4,	1, 3, 2, 4,
4, 1–2–3,	2, 3, 1–4,	2, 1–4, 3,	1, 3, 4, 2,
1–2, 3–4,	2, 4, 1–3,	2, 1–3, 4,	1, 4, 2, 3,
1–3, 2–4,	3, 4, 1–2,	3, 1–2, 4,	1, 4, 3, 2,
1–4, 2–3,			

Table 3.2

Number of legs	*Number of sequences*	*Number of potentially statically stable sequences*
1	1	0
2	2	0
3	6	0
4	26	6
5	150	114
6	1082	1030
7	9366	9295
8	94586	94493
9	1091670	1091552
10	14174522	14174376

Stability

The final column in Table 3.2 lists the number of potentially statically stable sequences. A statically stable gait is one in which the centre of gravity is always contained within the area enclosed by the supporting legs. For static stability it is therefore necessary that three legs, at least, are always in contact with the ground. This in turn requires a statically stable walking machine to have at least four legs—three legs in support whilst the fourth is raised and advanced. Table 3.2 shows that there are six quadruped sequences that satisfy this requirement: in each of the six only one leg moves at a time. However, although all six provide tripod bases for support, it will be seen later that only three of them lead to gaits that always contain the centre of gravity within the area of support.

One-, two- and three-legged animals and machines cannot be statically stable. The one-legged machine, the pogo-stick, must be kept in continual motion if it is not to topple. Raibert and Sutherland (1983) describe a machine that hops on a single leg and moves like a kangaroo in a series of leaps. Basically it achieves dynamic stability by leaping in the direction in which it tips. Thus, although statically unstable, it can be made dynamically stable.

Napier (1967) describes human walking as a unique activity during which the body, step by step, teeters on the edge of catastrophe. The rhythmic forward movement of first one leg and then the other keeps us from falling flat on our faces. Like the pogo-stick, although statically unstable, we are dynamically stable. Several researchers (Vukobratovic, 1973; Ogo *et al.*, 1978) have attempted to reproduce this dynamic stability in biped walking machines, but their limited success has required resort to complex mathematics and powerful computers. It is interesting to note, however, the simple mechanical device, (Figure 3.10) used in toys, that uses

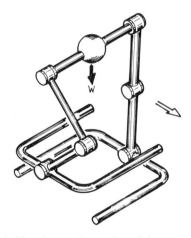

Figure 3.10   A two-legged walking mechanism.

big feet to ensure static stability at all times. The design ensures that the lateral extensions of each foot clear each other when a foot is being advanced.

Gaits

McGhee and Jain (1972) define gait as the characteristic pattern of footfalls. There is more to a gait than the sequence of footfalls that we have discussed earlier: the duration of each phase of the sequence must also be considered. Sequence and duration can both be accounted for it the gait is written in matrix form (McGhee, 1968). If the raised state of a foot is designated by a 1 and the placed state by an 0, then it is possible to construct a matrix that shows the state of all legs at all phases throughout the full cycle of a gait.

For example the gait matrix for a biped walk is constructed as follows:

	Leg 2	Leg 1
Phase 1	0	0
Phase 2	0	1
Phase 3	0	0
Phase 4	1	0

$$\equiv \begin{bmatrix} 0 & 0 \\ 0 & 1 \\ 0 & 0 \\ 1 & 0 \end{bmatrix}$$

At the commencement of phase 1 both legs are placed; at phase 2 leg 1 is raised; at phase 3 leg 1 is placed again so that both feet are on the ground; at phase 4 leg 2 is raised—and so back to phase 1 when leg 2 is placed again.

There are four events in a biped walk, two raisings and two placings, and hence there are four rows in the gait matrix. The fact that only one numeral changes in going from one row to the next shows that these events are distinct and separate in the walk. Consider, on the other hand, the gait matrix for the two-legged hop:

$$\begin{bmatrix} 0 & 0 \\ 1 & 1 \end{bmatrix}$$

Here raisings and placings occur simultaneously so that two numerals change in going from one row to the next. Such a gait is known as a singular gait (McGhee, 1968). Gaits such as the biped walk, in which simultaneous raisings or placings of feet do not occur, are called non-singular.

In general, it can be shown (McGhee, 1968) that there are $(2n-1)!$ non-singular gaits for an n-legged machine. For $n=2$, the six gaits are as follows:

$$\begin{bmatrix} 0 & 0 \\ 0 & 1 \\ 0 & 0 \\ 1 & 0 \end{bmatrix} \quad \begin{bmatrix} 0 & 1 \\ 1 & 1 \\ 1 & 0 \\ 1 & 1 \end{bmatrix} \quad \begin{bmatrix} 0 & 0 \\ 0 & 1 \\ 1 & 1 \\ 1 & 0 \end{bmatrix} \quad \begin{bmatrix} 0 & 0 \\ 1 & 0 \\ 1 & 1 \\ 0 & 1 \end{bmatrix} \quad \begin{bmatrix} 0 & 0 \\ 0 & 1 \\ 1 & 1 \\ 0 & 1 \end{bmatrix} \quad \begin{bmatrix} 0 & 0 \\ 1 & 0 \\ 1 & 1 \\ 1 & 0 \end{bmatrix}$$

(a)   (b)   (c)   (d)   (e)   (f)

Gait (a) is the biped walk and (b) the run in which there are instances when both feet are off the ground. Gaits (c) and (d) are sometimes used by children at play, especially when imitating galloping horses. Gaits (e) and (f) are not encountered in nature.

For quadrupeds, we saw earlier that there are 26 different possible sequences of leg operation and, of these, 20 involve simultaneous operation of two or more legs. Hence the remaining six sequences form the basis of the non-singular quadruped gaits. There are $7! = 5040$ non-singular quadruped gaits, each involving eight phases—four raisings and four placings.

For example the gait matrix for a horse's gallop, based on the sequence $(1, 3, 4, 2,)$ is (see Figure 3.11 for leg numbers):

$$\begin{bmatrix} 1 & 1 & 0 & 0 \\ 1 & 1 & 1 & 0 \\ 1 & 1 & 1 & 1 \\ 1 & 0 & 1 & 1 \\ 0 & 0 & 1 & 1 \\ 1 & 0 & 1 & 1 \\ 1 & 0 & 0 & 1 \\ 1 & 1 & 0 & 1 \end{bmatrix}$$

Leg 1 is placed at the start of phase 1, leg 3 at phase 4, leg 4 at phase 5 and leg 2 at phase 7. Note that there is always more than one leg raised at a time. Indeed at phase 3 the horse is in flight with all four legs off the ground.

The large number of non-singular gaits is reduced drastically if we demand, for purposes of static stability, that three legs are on the ground at any instant. This constraint reduces the number of quadruped gaits to six (known as creeping gaits) with the sequences:

(a) $1, 2, 4, 3,$	(b) $1, 3, 4, 2,$	(c) $1, 4, 2, 3,$
(d) $1, 2, 3, 4,$	(e) $1, 3, 2, 4,$	(f) $1, 4, 3, 2,$

For example the gait matrix for the quadruped crawl (sequence (c)) is:

$$\begin{array}{c}
\begin{bmatrix} 0 & 0 & 0 & 0 \\ 1 & 0 & 0 & 0 \\ 0 & 0 & 0 & 0 \\ 0 & 0 & 1 & 0 \\ 0 & 0 & 0 & 0 \\ 0 & 1 & 0 & 0 \\ 0 & 0 & 0 & 0 \\ 0 & 0 & 0 & 1 \end{bmatrix}
\begin{array}{l}
\text{time}/T \\
t_1 = 0 \\
t_4 - (1 - \beta) \\
t_4 \\
t_2 - (1 - \beta) \\
t_2 \\
t_3 - (1 - \beta) \\
t_3 \\
\beta
\end{array}
\end{array}$$

This shows that the quadruped crawl alternates between three-legged and four-legged stances.

The stability of a gait cannot be determined from the gait matrix alone; the time at which each phase commences must also be known. In the ensuing discussion we shall restrict ourselves to regular gaits in which all legs spend equal periods of time on the ground. The duty factor β is defined as the amount of time, as a proportion of the period T, that a leg spends on the ground. Hence in a regular gait all legs have equal duty factors.

t_j is the time normalized with respect to the period T, at which leg j is placed. Note that an interval $(1-\beta)T$ elapses between raising a foot and placing it again.

The discussion so far has been restricted to regular gaits. Let us now add a further restriction by requiring, as nature does, that the gaits be symmetric, i.e. the movement of a right leg (j)(odd) is shifted by half a period with respect to its partnering left leg $(j+1)$(even). Hence for the quadruped crawl

$$t_2 = t_1 + 0.5 \quad \text{and} \quad t_3 = t_4 + 0.5$$

Choosing $t_1 = 0$; $t_4 = 0.25$; $\beta = 11/12$ leads to the progression shown in Figure 3.11. In this diagram it is important to distinguish between step and stride. A step is here defined as the distance the animal or machine progresses whilst a given leg is in contact with the ground. A stride is the distance moved during one complete cycle of a particular gait. The relationship between step and stride can be determined as follows: If the machine proceeds at constant velocity V then, since each leg is on the ground for a time βT, the machine will progress by βTV whilst a foot is placed, i.e. the step S is βTV. The stride L, on the other hand is of length TV. Hence stride $L = S/\beta$.

Thus in Figure 3.11, the stride $= (12/11)$ times the step. The figure shows that leg 1 is placed at $t_1 = 0$. Leg 4 is then raised at $2T/12$ when the machine has moved forward by $L/6$. Leg 4 is placed again at $3T/12$ when the

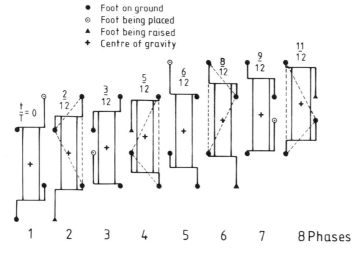

Figure 3.11 Progression of a four-legged walking machine (sequence 1, 4, 2, 3,).

machine has reached the point $L/4$, and so on. Support alternates between three and four legs and examination of the figure shows that the centre of gravity always lies within the base provided by the supporting legs. The machine is always statically stable during this gait. It has been shown (McGhee and Frank, 1968) that stability is optimal for this gait when $\beta = 11/12$, $t_1 = 0$, $t_4 = 5/12$, $t_2 = 1/2$, $t_3 = 11/12$ and, with the configuration of Figure 3.11, when the step is twice the axial distance from the centre of gravity to the leg fixing point.

For a regular creeping gait, requiring at least three legs on the ground at any time, the minimum value of the duty factor is 0.75. A gait with a lesser value of β would have instances with only two feet on the ground. Figure 3.12 shows the pattern of footfalls for all six of the quadruped creeps. Only phases 1, 3, 5 and 7 are shown, since for $\beta = 0.75$, the others, 2, 4, 6 and 8 when all four feet are placed, are of zero duration. The forward progression of the machine is not shown since, for the determination of stability, it is only necessary to know the position of the feet relative to the centre of gravity.

Inspection of the diagram shows that, for $\beta = 0.75$, the crawl sequence $(1, 4, 2, 3,)$ is the only stable gait, and even it is only marginally so at phases 1 and 3. However, Figure 3.11 shows how this gait could be stabilized by increasing β to 11/12. Lundstrom (1976) has shown that gaits $(1, 3, 4, 2,)$ and $(1, 2, 4, 3,)$ are stable if $\beta > 5/6$ but the other three gaits are only stable if $\beta = 1$, which is an unrealizable ideal.

Greater stability can be achieved by increasing the number of legs. For example, insects can move their six legs in 1082 different sequences (Bessonov and Umnov, 1976) and 1030 of these are potentially stable. However, the requirement for symmetry of motion reduces this to 24. Two of these are illustrated in Figure 3.13. The first (Figure 3.13(a)) shows the so-called tripod gait, in which three legs move at a time (1–4–5, 2–3–6,) and the machine always rests stably on a tripod base. This is, of course, a

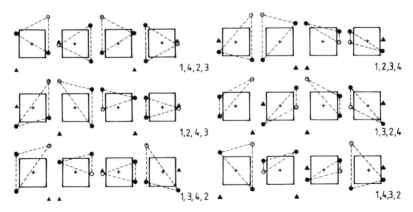

Figure 3.12   Footfall patterns for the six quadruped creeps.

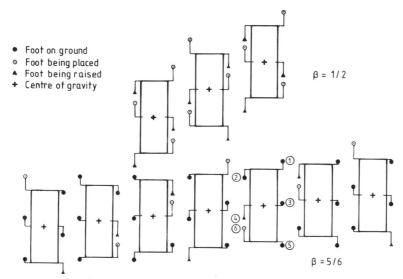

Figure 3.13 Two hexapod gaits: the tripod gait and the wave gait.

singular gait. Figure 3.13(b) illustrates a non-singular hexapod gait, based on the sequence $(2, 5, 3, 1, 6, 4,)$. This is known as a wave gait, i.e. a gait in which adjacent legs move successively. Such gaits are particularly stable. Reference to Figure 3.11, which showed a quadruped crawl, will confirm that it too was a wave gait $(3, 1, 4, 2,)$.

3.7 Energy consumed during walking

Nature favours two major types of leg configuration: the human type and the spider type. In the former the knee joint is situated beaneath the hip joint, whilst in the latter the knee joint is situated laterally to or higher than the hip joint. It has been argued (Hirose and Umetani, 1978) that the spider type leg offers the greater promise for off-the-road vehicles since it can extend its foot within a wide range, not only forwards and backwards, but also upwards and downwards about the hip joint. This allows it to circumvent obstacles easily. In addition, under similar conditions, the spider type leg consumes less energy than the human type.

  Let us calculate the energy consumed by a leg of the spider type. In so doing we must distinguish between useful work and dissipative work. When the force supplied by an actuator is in the same direction as its displacement, the work done is positive: this is called useful work. On the other hand, when force and displacement are in opposite directions the work done is negative and in many engineering systems this form of energy has to be dissipated. For example, in hydraulic systems this is done by forcing-high pressure oil back to tank across a relief valve.

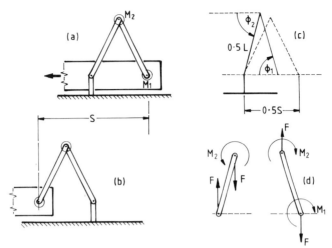

Figure 3.14 Forces and moments during a slow walk: (a) start of a spider leg step
(b) finish of step (c) geometry (d) static forces and torques.

Figure 3.14 shows a spider type leg with upper and lower legs of equal
length $0.5L$. The tarsal leg, carrying the foot, is assumed to stand vertical at
all instants. Rotary actuators at hip and knee generate torques M_1 and M_2
respectively. Figures 3.14(a) and (b) show the leg configuration at the start
and at the completion of a step of length S. The positive directions of the
angular variables ϕ_1 and ϕ_2 are defined in Figure 3.14(c). Figure 3.14(d) is a
free-body diagram showing the forces and moments acting on the upper and
lower legs. For the sake of simplicity, friction between foot and ground is
ignored: the ground reaction F is therefore vertical. Assuming constant
velocity and neglecting the mass of the legs, we can write

$$\text{lower leg:}\quad M_2 - 0.5FL\cos\phi_2 = 0 \tag{3.1}$$

where M_2 is the torque applied by the knee actuator.

$$\text{upper leg:}\quad M_1 + M_2 + 0.5FL\cos\phi_1 = 0 \tag{3.2}$$

where M_1 is the torque applied by the hip actuator. Now $\phi_1 = \phi_2$, hence
from (3.1)
$$M_2 = 0.5FL\cos\phi_2 \tag{3.3}$$
and from (3.2)
$$M_1 = -FL\cos\phi_1 \tag{3.4}$$

The energy expended in bringing the leg to the perpendicular is

$$\text{motor 2 (knee):}\quad E_2 = \int_{\phi_{20}}^{\pi/2} M_2\,d\phi_2 = 0.5FL(1 - \sin\phi_{20}) \tag{3.5}$$

where ϕ_{20} is the initial value of ϕ_2,

$$\text{motor 1 (hip):}\quad E_1 = \int_{\phi_{10}}^{\pi/2} M_1\,d\phi = -FL(1 - \sin\phi_{10}) \tag{36}$$

Thus the motor at the knee does useful positive work, whilst the motor at the hip does dissipative work while raising the leg to the vertical. This situation is reversed during the remainder of the step.

Since the power source has only to supply the total positive work over a cycle, then

$$E = 1.5FL(1 - \sin \phi_{20}) \tag{3.7}$$

where

$$\sin \phi_{20} = (1 - 0.25(S/L)^2)^{0.5} \tag{3.8}$$

Thus the energy requirement varies from a minimum of 0 at zero step length, to a maximum of $1.5FL$ at the maximum stride of $S = 2L$. For example, for a 1000 kg vehicle with $S = L = 1$ m, around 12.5 kJ of energy would be required to drive the legs. If the speed over the ground were 1 m/s then the motors would have to generate 12.5 kW, simply to support the body weight. Additional energy will be required to swing the legs.

In this particular mechanism the straight-line motion of the foot is achieved by two actuators working in opposition to each other. Hunt (1982) was one of the first to recognize that this could be avoided by using separate actuators for propulsion and for leg raising. This can be achieved by the use of orthogonal actuators (Figure 3.15). Note that for increased stability the feet are elongated in a direction at right angles to the direction of motion. As shown earlier (Figure 3.10) this device has been used to stabilize biped machines.

Two other methods of separating propulsion and leg raising are shown in Figure 3.16. In (a), with actuator A locked, the foot C is part of the coupler of the four-bar linkage $DEFG$ (Chapter 2), and for the given geometry part of the coupler curve for C is nearly a straight line (Hunt, 1982). Actuator B drives the machine forward and returns the foot and actuator A raises and places the foot.

In Figure 3.16(b) we again meet the pantograph mechanism discussed in Chapter 2. The horizontal and vertical actuators at B and A guide the foot around a rectangular path with respect to the vehicle (Taguchi *et al.*, 1976).

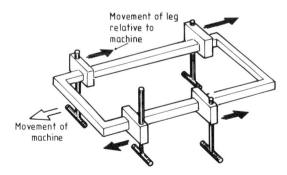

Figure 3.15 Separate actuators for propulsion and foot raising.

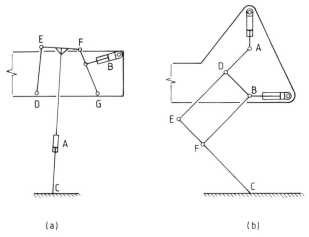

(a)   (b)

Figure 3.16 Other means of separating propulsion and foot raising: (a) using a four-bar linkage *DEFG* (b) using a pantograph.

However, in spite of their poor energy usage, it must not be forgotten that jointed legs are more manoeuvrable than the orthogonal variety and this may be of overriding importance in many applications.

3.8   Problems of control

Having decided on a particular design of leg and its actuation the next task is to choose a control system that will regulate the chosen gait. This is a relatively easy task if the terrain is flat and uniform. However, if there are obstacles present, an autonomous machine must have some means of detecting them and adjusting the step length and height accordingly. It would also need to ensure that in clearing such obstacles the machine is always in a statically stable configuration. If not, it may be necessary to circumvent, rather than clear, the obstacle, in which case the machine would also have to be able to steer iself. And, in addition, for purposes of comfort, it would be desirable to maintain the body in a constant orientation throughout such manoeuvres.

These control tasks are formidable and it is only now, with the power of the microprocessor at hand, that engineers have begun to tackle them with optimism (McGhee, 1978; McGhee and Iswahdhi, 1979; Bessonov and Umnov, 1983).

3.9   Recapitulation

We have seen that a robot has many anthropomorphic characteristics. Like a human, it needs wrists, hands and fingers in order to manipulate objects. There are also applications where the robot needs legs for mobility.

A wrist requires three degrees of angular freedom; pitch, yaw and roll. Several mechanical arrangements of the three drive motors are possible, each with its advantages and disadvantages.

Grippers fall into three major categories, based on mechanical linkages, magnets and vacuum. Whilst these are designed to lift objects of fairly well defined geometry, universal grippers and, ultimately, mechanical hands are much more versatile.

Mobility can provide a robot with an additional three degrees of freedom. We have argued that the wheel, with its many advantages, is limited to relatively flat surfaces; whereas rimless wheels, Venetian wheels, tracks and walking machines are capable of traversing rough terrain. Walking machines are a particularly exciting development but they present many difficult problems, including stability, choice of gait, control and guidance. Energy consumption is another possible problem.

Chapter 4

Actuating systems

4.1   Introduction

Actuating systems provide a robot with muscle power. They are energy conversion devices, converting electrical, hydraulic or pneumatic power to mechanical power. The basic drive elements may be classified into motors and actuators, the former capable of continuous rotation and the latter limited in motion, whether linear or rotary.

Hydraulics

A major advantage of hydraulic actuation is the ability to generate very large forces, made possible by the high working pressures of today's hydraulic systems. Pressures up to $280 \times 10^5 \, N/m^2$ (280 bar) are not uncommon, and even with an actuator piston of only 2 cm diameter such a pressure gives rise to a force of 8800 N. Some of the largest force applications are found in the metal-processing industry, where forges can apply forces up to 3 MN.

A high force-to-weight ratio is another important advantage of hydraulic actuation and this is particularly attractive in situations where weight is at a premium—such as aircraft, missiles and robots. Hydraulic motors are much smaller than electric motors with the same power rating (Figure 4.1(a)), the differences in size and weight again being attributable in the main to the high working pressure of the hydraulic system. It was mentioned earlier that 280 bar is not uncommon in hydraulic systems: by comparison the magnet in an electric motor can only exert a pressure of around 17 bar. This size effect can be explained by reference to Figure

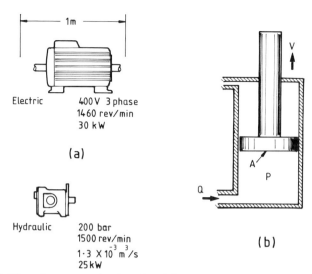

Figure 4.1 (a) Electrics versus hydraulics (b) hydraulic cylinder supplied with oil at
rate Q and pressure P.

4.1(b), which shows an hydraulic cylinder supplied with oil at a rate of
Q m³/s, and at pressure P N/m². The piston, of area A m², moves at velocity
V m/s. We can relate power to pressure and flow rate as follows:

$$\text{power} = \text{force} \times \text{velocity}$$
$$= (PA) \times V$$
$$= P \times AV$$
$$= P \times Q$$

Thus for a given power, an hydraulic actuator working at high pressure
requires only a small flow rate of oil; it can therefore be quite small.

The stiffness of an hydraulic drive is another important advantage. Oil is
relatively incompressible, making hydraulic drives relatively insensitive to
load disturbances. This is another reason for their popularity in the
aerospace and machine-tool industries. High stiffness also allows accurate
control to be achieved.

These advantages of high force, high force/weight ratio, high stiffness
and ease of control are offset to some extent by a number of practical
problems, the worst being:

• Hydraulic systems are expensive: manufacturing precision has to be high
  to maintain very small clearances between fixed and moving parts. This
  is necessary if leakage (and power wastage) is to be minimized.
• Since there is a possibility of leakage, it may be unwise to use hydraulic
  systems in environments where hygiene is important, i.e. food
  processing.
• Space has to be provided for the pipework.

Pneumatics

The earliest pneumatic systems (see McCloy and Martin (1980)), used air as
the working medium, but nowadays inert gases and hot gases are used in
some applications. Working pressures are usually limited to 7.0 bar, although at present some systems are being designed for 10 bar operation.
Inefficiency of compression of gases and the dangers involved in storing
high-pressure gases necessitate these limits.

The use of compressed air as a source of power has grown rapidly over
the past 20 years, and at present is accepted by all branches of industry. Its
advantages, many of which are shared by hydraulics, can be summarized as
follows:

- Most production plants have compressed air supply on hand.
- Pneumatic components are relatively cheap.
- Components are readily available 'off the shelf'. The range of valves is
  extensive and there is a wide choice of cylinder sizes.
- Components are reliable and are easily and cheaply maintained.
  Servicing can usually be carried out on site since it only involves
  replacing valve seals or cylinder bearings.
- Pneumatic actuators do not burn out when stalled. Hydraulic actuators
  also have this advantage.
- There is no fire hazard when pneumatics are used. They can be used for
  instance in situations where risk of explosion would prohibit the use of
  electricity.
- Compressed air is not inflammable, and hence pneumatic equipment can
  be used in high-temperature conditions where the use of hydraulics or
  electrics could be dangerous or costly. Most equipment can work up to
  80–90°C, but higher temperatures can be tolerated if heat-resistant seals
  are used.
- Pneumatic systems are clean.

Of course there are also associated disadvantages, and it is necessary to
consider these carefully before installing pneumatic equipment:

- Compressed air is expensive to produce, and this must be borne in mind
  if the plant does not already have a compressor. On a power basis it is
  considerably more expensive than electricity or hydraulics.
- Extreme accuracy of feed is difficult to achieve on account of the elastic
  nature of compressed air. It cannot compete with hydraulics or electrics
  in this respect.
- Transmission of air signals in pipes is much slower than the transmission
  of electrical signals in wires. Hence if signalling times are critical and
  lines are long (greater than 10 m), electrics should be used.
- Pneumatic cylinders are bulky and expensive if large forces are desired.
  Large forces are generated more conveniently by using hydraulics.
- Compressed air systems can be noisy.

Electrics

Most people are more familiar with electric power than with fluid power, since the former is common in every household and in most walks of life. With respect to robot applications, electric power offers several advantages (Kafrissen and Stephans (1984)):

Electric actuators are easy to control: they can provide fast accurate control of position and speed.
They are readily available and inexpensive.
It is easier to design for wiring than for piping.
Electric actuators are quiet in operation.
Electrical systems are clean.

But again, as in all areas of technology, there are disadvantages to be considered:

Power/weight and torque/weight ratios are low.
Low torques require extensive gearing and because of backlash this can introduce control problems.
Arcing can cause a fire hazard: electric actuators are therefore eliminated from certain tasks such as paint-spraying.
The possibility of electric shock introduces a safety hazard.

Which system is best?

The above properties of hydraulic, pneumatic and electric systems allow us to draw some conclusions concerning the potential use of each form of actuator in the robot field. Firstly it is necessary to distinguish between direct and indirect drives (Ray, 1983). Direct drives have no mechanical linkages between the actuator and the driven link. Hydraulic and pneumatic cylinders and motors can be used in the direct-drive mode because of their high force/torque capabilities. Direct drives have several advantages: they are compact, allowing them to be installed at the robot joints: they are simple and easy to maintain.

Indirect drives require a mechanical linkage between the actuator and the driven element, usually for the purpose of increasing the force/torque output. The linkages can take the form of gears, ball screws, harmonic drives, belts, chains, etc. The gearing ratio of indirect drives is usually in the range of 50:1 to 100:1 and this gives a stiff system, one that cannot easily be back-driven. This is a desirable feature in machining applications and in applications requiring fast movements over short distances. Fluid power direct drives exhibit a measure of compliance, particularly pneumatic systems, and this can be a disadvantage in such applications. Indirect drives allow the use of DC servomotors, which are quiet and efficient.

A major disadvantage of the indirect drive is the bulk and expense of

the associated mechanical linkages. In addition, backlash within these
linkages can affect repeatability and stability. Since the linkages are designed
to gear down the electric motors, the indirect drives are rarely suitable for
large robots requiring high-speed motions. Such robots generally require
direct hydraulic drives, whilst indirect electric drives will continue to be
popular for the smaller robots.

We now take a closer look at fluid power and electric drives: first, fluid
power systems.

4.2 Converting fluid power to mechanical power

Assuming that a fluid power source is available, the designer of a robot
system has the task of devising a means of controlling that power and of
converting it to mechanical power. We shall consider the latter task first.
Motors and actuators perform the opposite function to pumps and compres-
sors, in that they reconvert fluid energy to mechanical energy so that useful
work can be done.

Many hydraulic pumps can also be run as motors. Gear, vane and
piston motors are the commonest types on the market today (Figure 4.2). In
its simplest form the hydraulic gear motor consists of two meshing spur
gears, usually with eight or ten teeth, enclosed in a closely fitting housing.
The incoming high-pressure oil exerts a turning moment on both gear
wheels, although power is usually extracted from only one of them. The oil
is carried around between the teeth and discharged at the tank side. Gear
motors can operate at pressures up to 200 bar and can develop powers up to
100 kW.

An hydraulic vane motor is illustrated in Figure 4.2(b). The cylindrical
slotted rotor contains spring-loaded vanes which project radially. They are

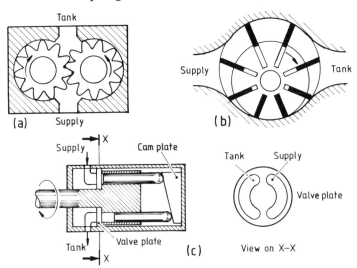

Figure 4.2 Fluid power motors: (a) gear motor (b) vane motor (c) axial piston
motor.

free to move in the slots, and are held in contact with the casing by the combined effects of centrifugal action and spring force. The rotor is placed eccentrically in the casing and this allows the incoming pressurized oil to exert a net clockwise torque on the rotor. The oil is carried round in the spaces between the vanes to the discharge port. Pressures are usually limited to around 70 bar.

Piston motors usually employ several small pistons reciprocating at high speeds. They are normally used for the higher hydraulic pressures (>150 bar). There are many different types, including plunger motors, radial motors and axial motors, of which only the last is discussed here. The axial piston motor is so called because the pistons move along the axis of the cylinder barrel (Figure 4.2(c)). As the piston is driven outwards, its contact with the cam plate forces the cylinder barrel to rotate. The valve plate, with its kidney-shaped ports, is fixed to the body of the motor.

There is also a wide variety of pneumatic motors with output powers in the range 0–20 kW. Piston motors are usually preferred for industrial applications where a good low-speed torque is essential. Vane air motors are popular for medium-power duties (<12 kW) and are used in portable tools, hoists, conveyor-belt drives, etc. Very high speeds can be achieved by the small turbine motors. For example pencil-type grinders can be driven at 80 000 rev/min by these motors, and dental drills can go up to 700 000 rev/min.

The simplest linear actuator is the single-acting cylinder or ram in which the fluid drives the rod in one direction only (Figure 4.3(a)). In some cases the retraction stroke may rely on the load or on gravity, and in others a spring may be used (Figure 4.3(b)). Since the cylinder has to do work against the spring, it is usual to restrict the use of spring returns to short

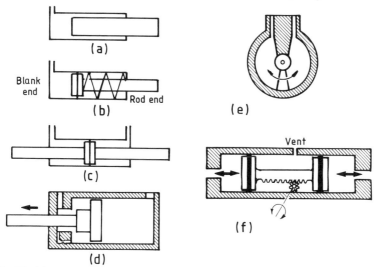

Figure 4.3 Fluid power actuators: (a) ram (b) spring-returned single-acting cylinder (c) double-acting cylinder (d) cushioning (e) vane-type rotary actuator (f) rack and pinion.

stroke applications such as clamping. In some hydraulic servomechanisms the return force is provided by means of a constant hydraulic pressure applied to the rod end.

If positive control is required during both extension and retraction of the piston rod, then it is necessary to use a double-acting cylinder (Figure 4.3(c)). This is usually driven by a valve which can raise the pressure on one side of the piston whilst simultaneously lowering it on the other side and vice versa. The cylinder in Figure 4.3(c) is a double-rod cylinder whose installation requires a space about three times its stroke. A single-rod cylinder, like that of Figure 4.3(b), can also be operated in a double-acting mode. It requires less space and is probably the most commonly used type. However, because of the different piston areas on each side, the forces and speeds developed during extension and retraction will be different, and as a result the double-rod cylinder tends to be the preferred arrangement for high-performance servosystems.

In pneumatic pick-and-place robots it is often necessary to include a deceleration cushion at the end of the stroke. This dissipates the kinetic energy of the load gently and reduces the possibility of mechanical damage to the cylinder, especially if fast-acting cylinders are used. Figure 4.3(d) illustrates the principle of operation. As the rod moves to the left the spigot enters the cushioning chamber and blocks the main escape path of the fluid. The fluid must force its way past the needle valve whose setting determines the degree of deceleration, and the kinetic energy of the load is dissipated as heat. A non-return valve is also necessary in double-acting cylinders so that, on reversal, the valve unseats, allowing unrestricted flow into the cylinder when the spigot is in the cushioning chamber.

Various external linkages can be employed to convert the linear output motion of a cylinder to a rotary output. There are, however, some actuators on the market that are specifically designed to give a restricted rotary output. Two are shown in Figure 4.3. The vane type (Figure 4.3(e)) can be either single- or double-acting, but its output rotation is restricted to about 300°. The rack and pinion arrangement of Figure 4.3(f) allows several rotations of the output shaft.

4.3   On–off control valves

These valves consist of two or more ports so arranged and connected that a controlling signal can *direct* the fluid to various parts of a system. Such valves are classified according to the number of ports they are designed to interconnect.

A three-port valve can control the movement of a single-acting cylinder (Figure 4.4(b)). In the normal position the cylinder is connected to exhaust and the spring holds the rod in the retracted position. Actuation of the valve pressurizes the cylinder and drives the rod to its extended position. This is the simplest of the on–off controllers. Its use can be extended to double-acting cylinders (Figure 4.4(c)). Two three-port valves are necessary in this

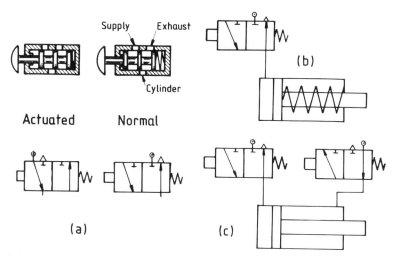

Figure 4.4 The three-port valve: (a) actuated and normal (b) controlling a single-acting cylinder (c) two valves controlling a double-acting cylinder.

case, so arranged that when one valve pressurizes one end of the cylinder, the other valve exhausts the other end.

One five-port valve can do the work of these two three-port valves. It alternately pressurizes or exhausts two cylinder ports. Figure 4.5(a) shows a five-port, two-position (5/2), pilot-operated, spring-return, pneumatic spool valve. Note that four lands are needed to ensure static pressure balance in both positions. Figure 4.5(b) shows the circuit for on–off control of a double-acting cylinder.

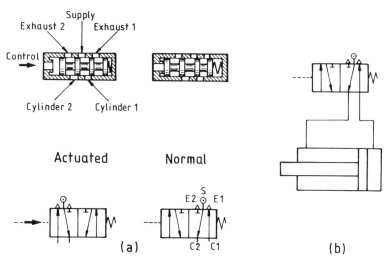

Figure 4.5 The five-port valve: (a) actuated and normal (b) controlling a double-acting cylinder.

4.4   Servovalves

The valves described above are used in applications where two-position control is adequate. Where accurate proportional control of position, velocity or force is demanded, a servovalve is essential. By definition a two-position valve controls two positions of a cylinder rod. Continuous control of position requires valves capable of an infinite number of positions: these are called servovalves. Figure 4.6 shows how three different types of servovalve could be used to drive a double-acting cylinder.

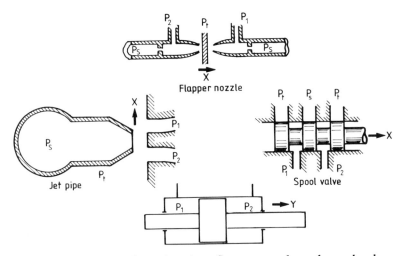

Figure 4.6   Servovalves: jet pipe, flapper nozzle and spool valve.

In the flapper nozzle valve a positive flapper movement x will reduce the exit orifice of the right-hand nozzle and increase that of the left-hand nozzle. Pressure P_1 will then rise and P_2 will fall. Flow will pass from the nozzle's P_1 port to the cylinder's P_1 port, and from the cylinder's P_2 port to the nozzle's P_2 port and hence to tank P_t.

The spool valve and the flapper nozzle valve illustrate respectively the use of shearing and seating elements. The jet pipe introduces a new type of valve element whose action relies on the conversion of the dynamic pressure of a jet into static pressure in an actuator. As the jet pipe is moved in the positive x-direction pressure P_1 increases and P_2 decreases.

Each of these servovalves is capable of driving the double-acting cylinder. There is a major difference, however, between the performance of the spool valve and the other two: the spool valve is close-centred, i.e. its cylinder ports are blocked when the valve is in mid-position. The flapper nozzle and the jet pipe valves are open-centred and cannot therefore hold a load quite so rigidly at a given position. And, unlike the spool valves, there is a steady flow of fluid in the mid-position, representing a quiescent power loss. This tends to restrict their use to low-power applications—usually with air. They have some advantages however, including a self-cleaning tendency, and relatively low cost.

4.5 A mathematical model

A mathematical model can provide a useful insight into the behaviour of an hydraulic actuator. With reference to Figure 4.7 we can establish the following approximate equation:

inlet flow rate through valve

$$Q_1 = k_1 x - k_2 P_1 \qquad (4.1)$$

where k_1 and k_2 are linearized valve coefficients. In reality Q_1 is non-linear in x (except when rectangular ports are used), and in P_1. However, a linearized analysis can give useful information. Assuming the valve to be symmetrical, the outflow rate through valve is

$$Q_2 = k_1 x + k_2 P_2 \qquad (4.2)$$

Now, with respect to the left-hand side of the cylinder, we can write

$$Q_1 = A\frac{dy}{dt} + C\frac{dP_1}{dt} \qquad (4.3)$$

or

inflow = displaced flow + compressibility flow

where A is the piston area and C is a compressibility factor. The

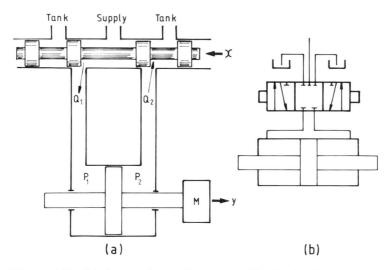

Figure 4.7 (a) Servovalve and actuator (b) circuit diagram.

compressibility flow arises because the fluid reduces in volume under pressure. Similarly,

$$Q_2 = A\frac{dy}{dt} - C\frac{dP_2}{dt} \tag{4.4}$$

Finally, if the load is a mass M, Newton's second law gives

$$A(P_1 - P_2) = M\frac{d^2y}{dt^2} \tag{4.5}$$

These equations combine to give the transfer function

$$\frac{y}{x}(s) = \frac{2k_1A/CM}{s(s^2 + 2\zeta\omega_n s + \omega_n^2)} \tag{4.6}$$

which indicates a potentially oscillatory response to step inputs. The natural frequency is

$$\omega_n = A\sqrt{2/MC} \tag{4.7}$$

and the damping ratio is

$$\zeta = k_2/(2A\sqrt{2C/M}) \tag{4.8}$$

Hence the natural frequency of such a drive is a maximum when the piston area A is high, when the load mass is low, and when the compressibility of the fluid C is low. The valve coefficient k_2 has a dominant effect on the damping ratio.

4.6   Hydraulic servomechanisms

Fluid power, particularly hydraulics, is widely used in position-control devices or servomechanisms. A typical hydraulic servo is illustrated in Figure 4.8. The output y follows the input x, but with force amplification. In a copy lathe, for example, the input x would be generated by a template follower, and the output y would be the tool position for reproducing the template's shape.

The electrohydraulic servo shown in Figure 4.8(b) is more common in robot systems, the amplifier A providing an interface between the computer and the rest of the system. In this case the valve consists of two stages, the pilot stage being driven by an electrical torque motor M. Movements of the pilot spool vary the differential pressure acting on the ends of the main spool, and this provides driving forces sufficiently large to move the main spool quickly against the opposing flow forces which tend to close the valve. Jet pipes and flapper nozzles are also used as pilot stages. Accurate position control, both of the load and of the main spool, requires negative feedback loops carrying positional information from the potentiometers on the main spool and on the cylinder rod. These signals are compared to the demand signal V_i generated by the computer.

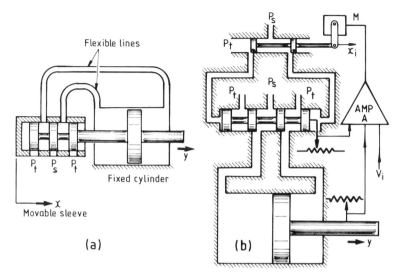

Figure 4.8 Hydraulic servos: (a) mechanical feedback (b) electrical feedback.

4.7 Electrical systems

Two major types of electric motors are used in robots: stepping motors and DC servomotors. Stepping motors find wide application in all computer-controlled machinery since a computer is ideal for generating the required pulsed control signals. Stepping motors move through a fixed angle for each pulse, and since the step size is fixed, a particular position can be demanded by sending the appropriate number of pulses to the motor.

Larger systems need high power and position measurement to facilitate feedback control, and here the DC motor predominates. It is an analogue device, and it will be shown that the torque generated is approximately proportional to the applied current.

Stepping motors

A robot requires programmable position control, and many smaller robots use stepping motors for this purpose. The stepping motor is particularly popular in small microcomputer-controlled robots, mainly because its use eliminates the need to monitor feedback signals. When pulsed in the correct manner, stepping motors rotate by a known amount, and by keeping count of the pulses it is possible for the microcomputer, knowing the geometry of the manipulator and gearing of the motors, to calculate the position of the robot. By switching stator windings in defined sequences, this type of motor produces an angular movement of fixed amount. Typical step sizes are 7.5°, 5° and 1.5°.

The method of operation of a unipolar stepping motor is illustrated in

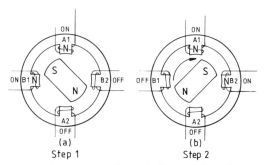

Figure 4.9 Unipolar stepping motor: (a) windings A1 and B1 are ON, A2 and B2 are OFF (b) windings A1 and B2 are ON, A2 and B1 are OFF.

Figure 4.9. There are two sets of phase windings, *A* and *B,* each winding being a north pole (N) when energized. In Figure 4.9(a) the south pole (S) of the permanent magnet attached to the rotor is attracted to windings *A*1 and *B*1 which are on. Switching *B*1 off and *B*2 on attracts the S pole to a new position 90° clockwise, as in Figure 4.9(b). Continuation of this switching sequence causes the rotor to step 90° clockwise each time, and reversal of the sequence produces anticlockwise movement.

The bipolar four-phase stepping motor shown in Figure 4.10 has a 30° step. The rotor has three N poles and three S poles, and the windings are energized as N or S poles in the sequence shown.

For even smaller step sizes it is necessary to increase the number of poles. An inherent feature of the stepping motor is its ability to maintain a static holding torque when energized.

Variable reluctance stepping motors can greatly increase the number of stable positions, i.e. reduce the step size. This may be achieved by stacking a number of stepping motors, with the stators aligned but the rotors spaced evenly. By this means three motors, for example, with step sizes of 7.5°, when combined, exhibit a stable position every 1.5°. The effect may be combined in one motor by having a different number of poles in the stator

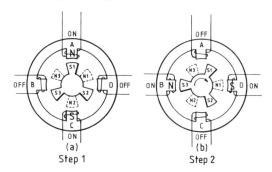

Figure 4.10 Bipolar four-phase stepping motor: (a) winding A is ON, attracting S1; winding C is ON, attracting N2 (b) winding B is ON, attracting S3; winding D is ON, attracting N1.

and rotor. A complicated relationship exists between the switching of the stator poles and the stable positions, but this is achievable under computer control. An example quoted by Lhote *et al.* (1984) is of a motor with eight stator windings and 50 rotor poles; different combinations of winding energization achieve 200 stable positions.

Direct current motors

The operation of the DC motor relies on the fact that a conductor in a magnetic field at right angles to it experiences a force perpendicular to the current and the magnetic lines of force. The magnetic field is created by permanent magnets or field windings in the stator (Figure 4.11(a)) and is radial. The rotor carries the armature windings, which lie axially, so that the force created when current flows through the armature causes the rotor to turn. The stator can have any number of poles greater than two; four are shown in the diagram. As each armature winding passes poles of alternating polarity, the direction of current needs to be changed so that the torque imparted to the rotor is always in the same direction. This is achieved by taking the current from brushes to the windings via commutator connections. Carbon brushes bearing on copper commutator segments ensure low friction and good electrical connection.

A DC motor may be controlled, i.e. its speed or torque varied, by altering the field or armature current. It is more usual to vary the armature current with the field current fixed, maintaining a constant magnetic field in the motor.

A mathematical model of such a DC motor can be constructed. Since the field is constant, the torque T exerted by the motor is proportional to the armature current i, i.e. $T = C_t i$.

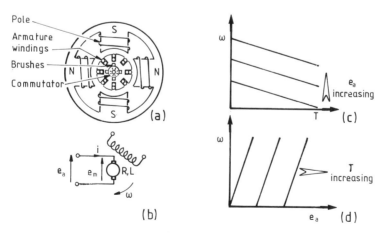

Figure 4.11 Direct-current motor: (a) four-pole stator with radial field and axial armature windings (b) electrical representation (c) constant voltage curves of speed against torque (d) constant torque curves of speed against voltage.

Rotation of the armature itself creates a field opposing the passage of current in the windings, and this 'back e.m.f.' e_m is proportional to the speed ω of the motor, i.e. $e_m = C_b\omega$. The actual voltage applied to the armature is therefore $e_a - e_m$, where e_a is the voltage applied to the brushes. This net voltage drives a current i through the armature against the resistance R and inductance L of the windings. Thus a model of the armature-controlled DC motor is

$$T = C_t i \qquad Ri + L\frac{di}{dt} = e_a - e_m \qquad e_m = C_b\omega \qquad (4.9)$$

These equations, together with dynamic considerations of inertial and friction loads (from bearings and brushes), will be used in Chapter 5 to develop laws for controlling the position of the motor by varying the voltage e_a.

Figures 4.11(c) and (d) show the steady-state characteristics of a DC motor. An increase in e_a is required to maintain a given speed with increased load, or to raise the speed at constant load. These motors are relatively high-speed/low-torque devices, and usually require reduction gearing in robot application. As indicated earlier, an electrical motor is heavier than an hydraulic actuator for a given power, and the provision of gearing exacerbates this situation. The solution in robots is often to mount the motor and gearing on or near the base and transmit the low speed/high torque required for a manipulator joint by mechanized means; some of these systems are discussed later.

Since the purpose of a DC motor is to convert electrical energy into mechanical energy, its efficiency is defined as the percentage of electrical power realized as mechanical power. An efficiency less than 100% indicates losses, and in DC motors these may be classified as load-sensitive and speed-sensitive (ECC, 1972). All losses are dissipated ultimately as heat, requiring cooling air to be introduced by a fan connected directly to the rotor shaft.

Load-sensitive losses are caused by current flowing through the DC resistance of the motor. The losses are load-sensitive since the current itself depends on load.

Speed sensitive losses can have various causes, for example:

- The resistance of the sliding connection between the brush and commutator, upon which a film of carbon, necessary for lubrication, has built up.
- Eddy currents and hysteresis in the armature core, arising from the changing magnetic field.
- Bearing and brush friction; the selection of brush material, size and pressure is important in a motor. As the pressure is raised friction increases and electrical resistance lowers, and an optimum pressure exists where the total loss is at a minimum (Figure 4.12(a)).
- Short circuits occur when a brush bridges two commutator segments

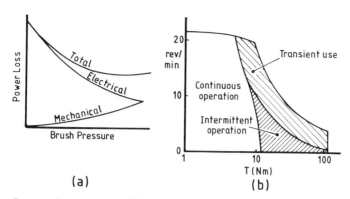

Figure 4.12 Losses in a motor: (a) relationship of friction and electrical losses to brush pressure (b) operating constraints on torque and speed.

momentarily, and any voltage differences cause current to flow, dissipating energy but performing no useful work.

A knowledge of losses allows the prediction of torques and speeds which may be demanded from a motor without overheating occurring. Since the heating effect is not instantaneous, intermittent demands can be imposed that would not be acceptable in continuous operation. Figure 4.12(b) shows a typical example of the constraints on the operation of one model (Morganite, 1961).

Current is consumed when a motor is working against a load, but on deceleration or with an overrunning load current could be recovered from the motor, when it is effectively functioning as a generator. This regeneration is not warranted on most robots, but may be justified on mobile machines relying on a limited battery capacity.

Disc and pancake motors

The conventional DC motor is widely used for robots, but other configurations are now finding increasing applications in robotics.

Figure 4.13(a) illustrates the principle of the disc motor, in which the

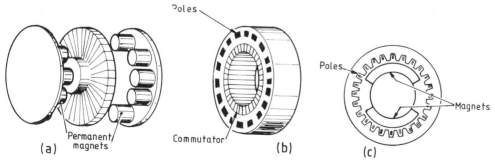

Figure 4.13 Direct current motor: (a) disc type (b) pancake type (c) brushless type.

armature carries current through radial conductors. The magnetic field is axial, created by permanent magnets on each side of the disc, so the force imparted to the disc causes it to revolve. The low armature inertia makes the use of disc motors attractive in applications where high accelerations are required.

The pancake motor, so called because of its high diameter/length ratio, is characterized by high torque at low speeds, making it suitable for direct drives (Margrain, 1983). The motor's many poles create a radial magnetic field, and the commutator has a large diameter with brushes making contact on the flat face (Figure 4.13(b)). Weight is saved and construction simplified by designing the motor and bearing as one unit.

Brushless DC motors

Instead of making the commutator windings rotate, it is possible to place the windings in the stator and to allow permanent magnet poles to rotate. No electrical connection is necessary to the rotor, so a brushless DC motor results (Figure 4.13(c)). The field must be revolved in order to make the rotor move, and this is effected by electronically switching the poles created by the stator windings. A large number of poles can give close positioning resolution; one machine uses a motor with 480 poles under computer control (Braggins, 1984). Very few brushless DC motors are found in robots at present, but research in this area is active (see for example Horner and Lacey, 1983) and they may be expected to play a greater role in the future.

Motors for direct drive

For direct drive, suitable motors are characterized by high torque at low speed. A direct-drive arm has been developed (Asada *et al.*, 1983) using DC motors operating on conventional principles. High torques are achieved using large-diameter motors; the largest is 560 mm in diameter and gives 204 Nm torque. Weight is reduced by using magnets made of samarium cobalt, an alloy which can have up to 10 times the magnetic energy of conventional alnico or iron magnets. Thus all the requirements for direct drive are met: light weight, high torque and low inertia for rapid acceleration.

4.8 Mechanical transmission

In Section 4.1 we saw that a direct drive has no mechanical linkage between a motor or actuator and the driven robot link. Indirect drive requires mechanical linkages for power transmission. It often involves a remote power source and may require mechanical transmission over a considerable distance. However, the term 'indirect' does not imply that the actuator cannot be placed at the joint. Self-contained units of motor, reduction gearing and position and velocity measurement devices are available for robots, and mounting these at the joints has been termed quasi-direct drive (QDD) (Hartley, 1984).

Remote drives

The construction of most manipulators requires the shoulder actuator to move the load on the gripper, plus the arms and bearings of the machine, plus the elbow and wrist actuators. If actuators can be placed nearer the base there is a two-fold gain: the load on the shoulder is reduced and deflection of the arms is reduced. An example of this technique was illustrated previously (Figure 3.2(c)) where the three rotations in the wrist were achieved by concentric drives in the arm, the motors being placed at the elbow, thus reducing the load on the elbow and shoulder actuators. It is relevant here to mention three more devices which may be used to drive a joint remotely: cables, belts and chains; linkage mechanisms; and ball screws.

Chain drives are typically found in large SCARA manipulators where the elbow actuator is mounted at the base (Figure 4.14(a)). Smaller machines may use belt drives (Figure 4.14(b)) or cable drives (Figure 4.14(c)). Even with a tensioning mechanism, all of these devices are subject to backlash due to stretching of the drive medium. Chains and toothed belts are positive drives, but a cable can slip over prolonged operation.

The cable drives shown in Figure 4.14(c) pitches the elbow under control of a motor mounted in the base. The pulley at the elbow is free to rotate, so that the angle of the elbow does not affect the pitch angle of the wrist. In order to move the end-effector of a robot in a useful path, e.g. a

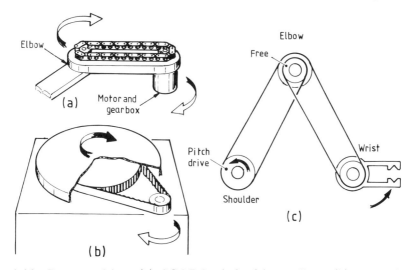

Figure 4.14   Remote drive: (a) SCARA chain-driven elbow (b) rotary-base belt drive (c) cable-driven pitch.

straight horizontal line, complex and coordinated movements are required
from the various actuators. The wrist drive of Figure 4.14(c) is one way of
reducing these interactions, so that driving one actuator affects only one
world coordinate. The Locoman (Pham, 1983), developed at Birmingham
University and marketed by Pendar Robotics, is a notable example of a
linkage mechanism designed to remove the interaction in a jointed-arm
configuration. Three motors in the base separately drive the end-effector in
the three world coordinate axes. The manipulator uses a combination of
pantographs and parallelograms (Figure 4.15(a)), so that linear motions in
world coordinates in the base are reflected as movements along the same
axes at the end-effector. Controlled path movements of the tool are thus
achieved without the need for transformation calculations.

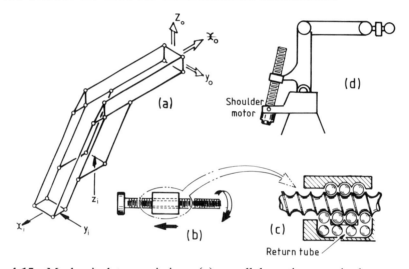

Figure 4.15 Mechanical transmission: (a) parallel motion manipulator configuration (b) linear motion from ball screw (c) ball screw construction (d) rotating
shoulder driven from ball screw.

Stepping motors are used in this robot, and their rotary motion is
transformed to linear motion by ball screws. A ball screw consists of a
bearing with an internal thread located on a screwed rod. If the rod is
rotated and the bearing is constrained so as not to rotate, it will move
linearly; conversely, if the rod is fixed linear motion can be achieved by
rotating the bearing (Figure 4.15(b)). The bearing is not simply an internal
thread, but a screw-shaped ball bearing (Figure 4.15(c)). As the screw
rotates the balls move round the screw thread until they reach the end of the
device, where they enter a return tube. The full name of the device is the
recirculating ball screw. The ball screw is suitable for configurations where
linear motion is required, e.g. cylindrical or cartesian configurations. The
triangular geometry of Figure 4.15(d) is also widely used to produce rotary
motion on jointed-arm manipulators.

Gearing

In robots, gears may be used to alter the speed and/or direction of a rotary motion, and to change rotary motion to linear motion or vice versa.

An example of conversion from linear to rotary motion was shown in Figure 4.3(f), where a rack and pinion is used to take rotary motion from the linear movement of a piston rod. The rack and pinion may also be found in robots where a linear motion is desired from a motor (Figure 4.16(a)).

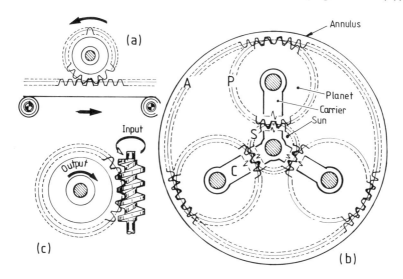

Figure 4.16 Gear transmission: (a) rack and pinion (b) epicyclic gear train (c) worm drive.

It will be seen in Chapter 6 that shaft speeds are often geared up in order to get good resolution from a position-measurement system. But in the actuation of robots the requirement is invariably for reduction gearing, since most motors are high-speed/low-torque. Spur gears are satisfactory for modest speed reductions; the reduction in speed is equal to the ratio of the teeth on the small wheel to those on the larger one.

The epicyclic gear train (Figure 4.16(b)) is more versatile and offers the advantage of in-line input and output shafts. There are four elements to the epicyclic transmission: sun wheel (S teeth), planet wheels (each with P teeth), held by a carrier (denoted by C) and an annulus, which has A internal teeth. The geometry of the gear train dictates that $A = S + 2P$, and $P < 0.464A$. The normal use of an epicyclic gear train is to fix one element, take a second element as input and a third as output. In theory, therefore, there are 24 ways of transmitting rotation through the train, but not all of these are practical. The input/output ratio for any combination can be determined as follows:

(1) Revolve the whole train one revolution clockwise.
(2) Hold the proposed output, and revolve the proposed fixed member one revolution anticlockwise.
(3) Add up the resulting rotations.

For example, the usual method of speed reduction is to fix the annulus, rotate the sun and take an output from a shaft attached to the carrier. Let $S = 20$; $P = 60$; $A = 140$:

	S	P	C	A
One revolution clockwise	$+1$	$+1$	$+1$	$+1$
Hold carrier, annulus one revolution anticlockwise	$+7$	$-2\frac{1}{3}$	0	-1
Sum	$+8$	$-1\frac{1}{3}$	$+1$	0

The resulting reduction is $8:1$.

Higher reductions are obtainable by building multiple epicyclic trains into one box, but this is undesirable in robots since there is an increased backlash which can cause positioning errors and instability in control systems. High reductions are also obtainable from worm drives (Figure 4.16(c)), which give a reduction of $N:1$ using a single start worm and a gear of N teeth. The worm drive is sometimes used to rotate the base of small robots where the arrangement is mechanically suitable and has the advantage of being self-locking, i.e. the output is locked even when the input is free from applied torque. A worm drive exhibits backlash which increases if axial play exists in the input bearings.

Robots often require large reduction boxes with no backlash; this cannot be provided in a light structure by conventional gearing, and to fulfil this need many manipulators use the harmonic drive.

The harmonic drive

The harmonic drive transmission system (Kafrissen and Stephans, 1984) is also a reduction gear, but its importance in robotics warrants a separate treatment here. The harmonic drive is a patented device manufactured by the Harmonic Drive Division of Emhart Machinery Group, Wakefield, USA. It is used when electric motors are placed at the robot joints. It gives a large reduction ratio; it is mechanically rigid, light and free of backlash.

The construction of a harmonic drive takes two forms, one of which is shown in Figure 4.17(a). The input shaft drives an elliptical 'wave generator', which is simply a ball bearing assembled round an elliptical former. The 'Flexspline', the output element, is a cup carrying an external gear. The wall of the cup is flexible; it is circular in the free state, but its flexibility allows it to conform to the circumference of the wave generator. The external teeth on the Flexspline engage with teeth on an internal gear, which is rigid and fixed to the body of the transmission.

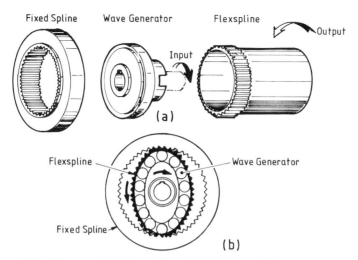

Figure 4.17 Harmonic drive: (a) exploded view (b) gear arrangement.

The teeth on both gears have the same pitch, but the Flexspline has fewer teeth than the internal gear. Figure 4.17(b) shows that the major axis of the ellipse formed by the teeth of the Flexspline is equal to the pitch diameter of the internal gear. The effect is exaggerated in the diagram in order to illustrate how the teeth mesh in two diametrically opposite areas; in practice about 10% of the teeth are always meshed. This is much higher than the typical spur gear.

If the internal gear has N teeth and the Flexspline F teeth ($F < N$), the gearing ratio is $(-N/F) + 1$. This is always negative, indicating that the output moves in the opposite direction to the input. $N - F$ must be an even number, since the gears must mesh at 180° intervals. Typically, $N = 202$ and $F = 200$, giving a gearing ratio of $-1/100$.

The harmonic drive is particularly suitable for large reduction ratios; 320 : 1 is available. Since $N - F$ is usually 2, the reduction ratio is therefore half the number of teeth on the Flexspline.

The reason for the popularity of the harmonic drive can now be identified. The drive is light since there are few parts in the assembly, and the gears may be sized in the knowledge that a good proportion of the teeth are meshed at any time; this aspect of the design also eliminates backlash, and contributes to torsional rigidity.

Another form of harmonic drive, pancake gearing, although exhibiting some backlash, is useful where space is at a premium. In this device the Flexspline is simply a flexible gear spline, with no cup body. The input turns a wave generator and there is a fixed rigid gear, as before, but the output is taken from another gear with internal teeth meshing with the Flexspline. This assembly is shorter in the axial direction since the cup is not needed to transmit the output rotation.

4.9   Recapitulation

Actuators for robots are devices to convert electrical, hydraulic or pneumatic power to mechanical power. The advantages of these three sources of
power and the principles of operation of the actuators have been outlined. A
distinction has been drawn between direct and indirect drive; hydraulic
actuators are suitable for direct drive, whereas electric motors often require
gearing to increase torque. Various forms of remote drive are used to
remove the weight of an actuator from a joint.

Chapter 5

Modelling and control

5.1 Introduction

The function of a robot is to perform useful tasks, and this cannot be done without a means of controlling the movements of the manipulator. Chapter 1 showed that in the simpler applications the position of an end-effector may be controlled by placing mechanical stops on the actuator. Such pick-and-place devices will be examined more closely in this chapter.

The relative inflexibility of the pick-and-place device precludes its use in the more demanding applications. For these it is essential to use closed-loop control, in which the error between a desired and an actual variable is used for corrective purposes. In many of these systems control of the end-effector's position is the major concern but, increasingly, applications require control of velocity and acceleration as well as position along a desired trajectory. The prescribed motion is maintained by applying corrective torques or forces to the actuators to adjust for any deviations of the arm from the trajectory. This requires a servomechanism on each axis of the robot to control the machine coordinate and its derivatives.

Control of a robot arm involves an extremely difficult analytical task. The dynamics of a robot with n degrees of freedom are highly non-linear: they are described by a set of n highly coupled, non-linear, second-order differential equations (Bejczy, 1974). The non-linearities arise from inertial loading, from coupling between adjacent links of the manipulator and from gravity loads—all of which vary with the machine coordinates. These difficulties are addressed later in the chapter.

In any control system it is important to know precisely the desired value of the controlled variable. Where robots are taught by pendant or lead-through, the desired values of each machine coordinate are recorded during teaching. In many applications however, teaching by doing is neither appropriate nor satisfactory. A growing number of applications use off-line programming, in whch the desired trajectory is expressed in world coordinates. The computer must then perform the inverse transformation from world coordinates to machine coordinates, in order to determine the desired values of the machine coordinates. Two methods of path control, geometric and kinematic, are considered later.

Since each axis of a robot requires a servomechanism is is relevant to include a brief review of continuous linear closed-loop control systems. Digital systems also receive a brief mention. They are particularly useful for optimizing the performance of the robot control system. Since the equations are non-linear, optimum control performance requires the parameters of the control system to be altered throughout the robot's working volume. Digital control techniques allow the control strategy to be determined by computer and are thus of great importance in robotics.

Finally, decoupled closed-loop control using rate controllers and acceleration controllers is introduced.

5.2   Sequence control

Much automatic machinery performs repetitive sequences of operations, and a variety of mechanical, pneumatic, fluidic and electronic devices can be used to generate such sequences. Sequences may be time-based, in which case actions are initiated at predetermined times, or event-based, in which case each step is initiated by the completion of the previous step.

Time-based sequence control

The camshaft programmer has traditionally been used to generate sequences for a variety of electric, hydraulic and pneumatic machinery.

The device consists of a series of cams mounted on a common shaft which rotates at a constant speed. Each cam directly operates and releases an electrical switch or fluid power valve.

Figure 5.1(a) shows a simple sequence of extensions (e) and retractions (r) over a 4 s cycle for a machine with three cylinders A, B and C, and Figure 5.1(b) shows the cams necessary to implement the sequence, with the shaft running at 0.25 rev/s.

Event-based sequence control

The cam programmer generates a particular sequence over a given period of time, but often it is necessary to ensure that each step of the sequence is completed before the next starts, i.e. event-based control. Much dedicated pneumatic machinery has this requirement, and event-based sequences can

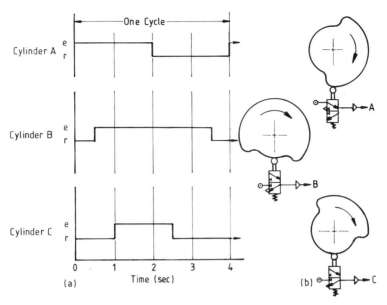

Figure 5.1 Time-based sequence: (a) timing of extensions, *e* and retractions, *r* (b) cams and associated valves.

easily be generated using pilot-operated valves. The cascade system (McCloy and Martin, 1980) may be used to design the pneumatic circuit; Figure 5.2 shows the cascade circuit for the sequence *Ae Be Ce Ar Cr Br.* (again *e* indicates an extension, and *r* a retraction). Completion of each extension and retraction in the sequence operates a limit valve, and the pilot signal from this valve then initiates the next step in the sequence, either by operating the appropriate main valve directly, or indirectly through a group selector valve. The selector valve is necessary to prevent a locked-on signal, a condition which occurs when a pilot signal cannot operate a valve due to the continued presence of an opposing pilot signal.

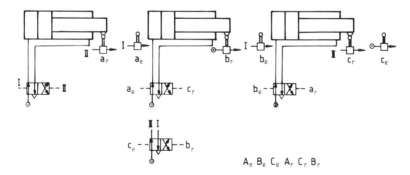

Figure 5.2 Event-based sequence: cascade circuit for *Ae Be Ce Ar Cr Br.*

Programmable logic controller (PLC)

The PLC is capable of providing both time-based and event-based control (C & I, 1986). Consider a pneumatic pick-and-place device with two degrees of freedom. Extension of cylinder A provides the gripping action, i.e. Ae = grip. Similarly Ar = release, Be = up, Br = down, Ce = out and Cr = in. The device is required to lift a bolt from an automatic feeder, set it on an indexing table, lift the gripper clear of the table, signal the table to index and return the gripper to acquire the next bolt. The required sequence, assuming all cylinders to be retracted at start, is

grasp a bolt	Ae
up	Be
out	Ce
down	Br
release the bolt	Ar
up	Be
turn on indexing signal for 0.5 s	
in	Cr
down	Br

Since event-based and time-based functions are both involved it would be difficult to implement this circuit using either cams or cascade circuits alone. In the case of a cascade circuit, the indexing signal could be derived from an air-operated electrical relay and timer. But this particular application requires cylinder B to extend and retract twice in each sequence, a requirement involving an excessive number of components in a cascade circuit.

The PLC dominates in this type of application. A basic PLC is a microprocessor-based device which switches a number of electrical outputs on and off, according to a preprogrammed sequence, including timing and decision-making based on the status of a number of electrical inputs. Manufacturers either offer a range of package models which may differ only in the number of inputs and outputs provided, or confine each function to individual modules, allowing the user to construct a PLC to suit his particular application. The basic functions will be evident from a front view of such a PLC (Figure 5.3) (Wickman, undated). From left to right, the first module houses the sequence controller itself, the lights on the panel displaying the status of all the inputs and outputs (in this case 16 inputs and 8 outputs). The microprocessor in this module controls the sequence, including any timing functions, according to the program which is stored in an erasable programmable read-only memory (EPROM). The EPROM does not lose the program when power is switched off.

This particular PLC also contains an amplifier which can accept analogue signals which could be from a transducer measuring, for example, temperature, pressure or torque. There is also a module which compares the

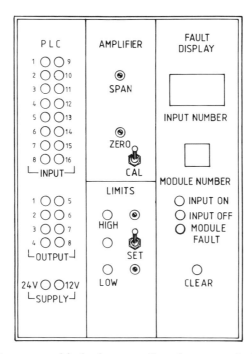

Figure 5.3 Programmable logic controller: front panel of typical model.

signal from the amplifier to preset limits and displays a 'high', 'within limits' or 'low' signal. This signal is transmitted to the common busbar at the rear of the unit, which connects all modules, where it may be interrogated during the execution of the program. If this is interrupted the fault display module on the right indicates which step of the sequence was being executed at the

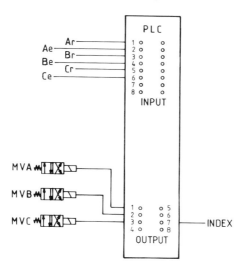

Figure 5.4 Programmable logic controller: connections for typical application.

time of interruption, and the number and status of the input being monitored at that point.

Figure 5.4 shows the equipment necessary to control any sequence of operations of the pick-and-place device mentioned. The pneumatic limit valves *Ar, Ae, Br, Be, Cr* and *Ce* have been replaced by simple microswitches which are connected to PLC inputs 1 to 6. The cylinders *A, B* and *C* are extended and retracted by operating or releasing solenoid-operated spring-return valves from outputs 1, 2 and 3. The indexing signal is taken from output 7. Figure 5.5 shows the flow chart of the PLC program for the above sequence. The first operation is to switch on the output connected to *Ae,* which causes a grasp movement. The input connected to the microswitch which is made when the gripper has closed is then interrogated: '*Ae* on?' If the answer is no (N) the program loops back to repeat the interrogation until the gripper has closed; the answer yes (Y) then initiates the next step *Be.* Once the flow chart has been compiled, the PLC program

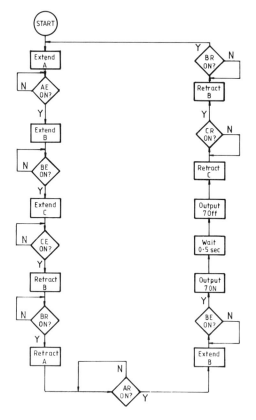

Figure 5.5 Flowchart of sequence involving time-based and event-based decisions.

is easily written, and the program for this particular sequence is:

```
10    OUTPUT 1 ON
20    WAIT FOR INPUT 2 TO COME ON
30    OUTPUT 2 ON
40    WAIT FOR INPUT 4 TO COME ON
50    OUTPUT 3 ON
60    WAIT FOR INPUT 6 TO COME ON
70    OUTPUT 2 OFF
80    WAIT FOR INPUT 3 TO COME ON
90    OUTPUT 1 OFF
100   WAIT FOR INPUT 1 TO COME ON
110   OUTPUT 2 ON
120   WAIT FOR INPUT 4 TO COME ON
130   OUTPUT 7 ON
140   WAIT 0.5 s
150   OUTPUT 7 OFF
160   OUTPUT 3 OFF
170   WAIT FOR INPUT 5 TO COME ON
180   OUTPUT 2 OFF
190   WAIT FOR INPUT 3 TO COME ON
200   GO TO 10
```

To enter the program into the PLC, it would be first typed into a microcomputer in the above format. the microcomputer translates the program into the language of the microprocessor in the PLC—machine language (see Chapter 7)—and transmits it to a programmer unit. The unit is capable of programming the PLC directly, making a tape or disc copy of a program and programming the PLC module from tape or disc.

Endstops

The pick-and-place unit described above would be inflexible without some form of control over the two extreme positions which each axis can adopt. This can be achieved by mechanical adjustment of the endstops against which a cylinder stalls at each end of its travel. Intermediate positions are often added to pick-and-place axes by additional on–off stops.

5.3   A dynamic model of a robot

A loaded manipulator is a dynamic system, and it is possible, knowing the desired trajectory of the end-effector, to express the forces or torques required from the actuators to achieve this motion in terms of the positions, velocities and accelerations along the path and the parameters of the manipulator. The expression may be derived by the energy approach, as exemplified by Lagrange (Paul, 1982), or the Newtonian approach, which relates forces and torques to masses, inertias and accelerations (Luh *et al.*, 1979). The equations stated in this section assume that the manipulator is made up of rigid links.

Dynamic equation of a robot

The dynamic control problem is, given the desired trajectory of the end-effector, what torque patterns (as functions of time) should be applied by the actuators to achieve the desired motion? Having a geometric model of the manipulator, the computer is able to transform the required path from world coordinates to machine coordinates, and having a dynamic model, which relates torques to position, velocity and acceleration of machine coordinates, allows the computer to predict the required torque patterns. The dynamic model takes the form

$$T + [W(\Theta)] = [M(\Theta)]\ddot{\Theta} + [N(\Theta)](\dot{\Theta})^2 + [P(\Theta)]\dot{\Theta}_i\dot{\Theta}_j \qquad (5.1)$$

where Θ is the vector of machine coordinates and all the matrices are in general non-linear. $[W(\Theta)]$ gives the torques applied due to gravity, and M, N and P quantify the torques arising from inertial, centrifugal and Coriolis forces respectively. If the machine has n degrees of freedom, M and N have dimensions $n \times n$. The Coriolis matrix P has dimensions $n \times C_n^2$ since these forces arise from combinations of the $\Theta_i\Theta_j$ product.

Calculation of dynamic coefficients

The main problem with the application of dynamic control in real time is the complexity of many of the terms in eqn (5.1). To illustrate this we consider the three-degrees-of-freedom manipulator shown in Figure 5.6, making the simplifying assumptions that the masses of the arms are concentrated at their centres, a concentrated mass m_{23} represents the elbow actuator and the

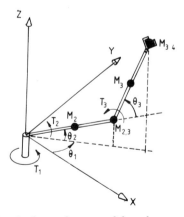

Figure 5.6   A three-degree-of-freedom manipulator.

wrist, gripper and load are represented by a concentrated mass m_{34}. Links 2 and 3 have respective lengths L_2 and L_3. The required torques are given by eqn (5.1) which in this case has the form

$$
\begin{bmatrix} T_1 \\ T_2 \\ T_3 \end{bmatrix} + \begin{bmatrix} W_1 \\ W_2 \\ W_3 \end{bmatrix} = [M(\Theta)] \begin{bmatrix} \ddot{\theta}_1 \\ \ddot{\theta}_2 \\ \ddot{\theta}_3 \end{bmatrix}
$$

$$
+ [N(\Theta)] \begin{bmatrix} (\dot{\theta}_1)^2 \\ (\dot{\theta}_2)^2 \\ (\dot{\theta}_3)^2 \end{bmatrix} + [P(\Theta)] \begin{bmatrix} \dot{\theta}_1\dot{\theta}_2 \\ \dot{\theta}_1\dot{\theta}_3 \\ \dot{\theta}_2\dot{\theta}_3 \end{bmatrix} \quad (5.2)
$$

The analysis is given in Appendix I, but the results are presented here to illustrate the complexity, the cross-coupling and the non-linearities involved. The presentation is simplified by identifying the following effective masses and inertias:

$$
\begin{aligned}
M_a &= 0.5m_2 + m_{23} + m_3 + m_{34} & M_b &= m_3 + 2m_{34} \\
I_a &= (0.25m_2 + m_{23} + m_3 + m_{34})L_2^2 \\
I_b &= (m_3 + 2m_{34})L_2L_3 & I_c &= (m_3 + 4m_{34})L_3^2
\end{aligned}
$$

The following shorthand notation is also useful:

$$
C(\alpha + \beta) = \cos(\alpha + \beta) \qquad S(\alpha + \beta) = \sin(\alpha + \beta)
$$

$$
\begin{aligned}
T_1 &= [I_aC^2\theta_2 + I_bC\theta_2C(\theta_2 + \theta_3) + 0.25I_cC^2(\theta_2 + \theta_3)]\ddot{\theta}_1 \\
&\quad - [I_aS(2\theta_2) + I_bS(2\theta_2 + \theta_3) + 0.25I_cS(2\theta_2 + 2\theta_3)]\dot{\theta}_1\dot{\theta}_2 \\
&\quad - [I_bC\theta_2S(\theta_2 + \theta_3) + 0.25I_cS(2\theta_2 + 2\theta_3)]\dot{\theta}_1\dot{\theta}_3 \\[4pt]
T_2 &- [M_aL_2C\theta_2 + 0.5M_bL_3C(\theta_2 + \theta_3)]g \\
&= [I_a + 0.25I_c + I_bC\theta_3]\ddot{\theta}_2 + [0.25I_c + 0.5I_bC\theta_3]\ddot{\theta}_3 \\
&\quad + [0.5I_aS(2\theta_2) + 0.125I_cS(2\theta_2 + 2\theta_3) \\
&\quad + 0.5I_bS(2\theta_2 + \theta_3)]\dot{\theta}_1^2 \\
&\quad - [0.5I_bS\theta_3]\dot{\theta}_3^2 - [I_bS\theta_3]\dot{\theta}_2\dot{\theta}_3 \\[4pt]
T_3 &- [0.5M_bL_3C(\theta_2 + \theta_3)]g \\
&= [0.25I_c + 0.5I_bC\theta_3]\ddot{\theta}_2 + [0.25I_c]\ddot{\theta}_3 \\
&\quad + [\{0.5I_bC\theta_2 + 0.25I_cC(\theta_2 + \theta_3)\}S(\theta_2 + \theta_3)]\dot{\theta}_1^2 \\
&\quad + [0.5I_bS\theta_3]\dot{\theta}_2^2.
\end{aligned}
\quad (5.3)
$$

Thus knowing a desired trajectory, and the velocities and accelerations along that trajectory, it is possible by inverse transformations to calculate the machine coordinates and their derivatives and hence to determine the actuator torques required to achieve the desired trajectory. This can be time-consuming, however: it has been demonstrated that if the Lagrange method is used, about 8 s of Fortran simulation is required to compute the move between two adjacent points on a planned trajectory for a six-joint robot (Luh *et al.*, 1980a). This is clearly too slow for on-line applications. Computation time can be reduced by simplifying the equations, for example, by neglecting the second-order Coriolis and centrifugal reaction terms, but such approximate models give suboptimal dynamic performance and require arm speeds to be restricted. Fortunately a method has been developed which is both fast and accurate. It relies on the Newton formulation of the dynamic equations, and its speed is due to the fact that the equations are recursive; i.e., the same equation can be applied to each link in turn. It is claimed (Lee, 1982) that with this method, only 3 ms (PDP11/45 computer) are required to calculate the joint torques for a given trajectory point.

5.4   Geometric and kinematic path control

The geometric and kinematic methods of path control assume that the manipulator passes through a succession of equilibrium states. As we have seen above this is likely to be a reasonable assumption only when powerful actuators are used and speeds are slow. Geometric control determines desired values of the machine coordinates whilst kinematic control determines desired values of the rate of change of the machine coordinates.

Geometric control

The geometric control outlined here is commonly used to generate motion along desired paths at controlled speeds. This is achieved by calculating a stream of machine coordinates in real time, representative of intermediate world coordinates along the desired trajectory. When the desired machine coordinates are determined the n servomechanisms (for n degrees of freedom) will drive the end-effector to the desired world coordinates through an unpredictable path. There will be errors in the actual coordinates achieved due to backlash in bearings, flexing of the links, and the resolutions of the position control systems.

   The inverse transform for a robot with six degrees of freedom can be a formidable task. Paul (1982) illustrates this for the Stanford Manipulator, and the Unimation PUMA is dealt with by Lee and Ziegler (1983). Here we consider a parallel-linkage robot (similar to the GEC Gadfly), basically because the task is less demanding of time and space. The arrangement is illustrated in Figure 5.7. The initial reference position is shown in Figure 5.7(a), with the equilateral plate holding the gripper disposed symmetrically at a height h above and parallel to the equilateral triangle formed by the earth fixing points. The world coordinate frame *XYZ* has its origin on the ground at the centre of the two concentric circles which circumscribe the plate 234 and the earth fixing points 567 (radii r and R respectively). Figure 5.7(b) is a plan view of the initial reference position in which the *xyz* coordinates of the gripper are (r, O, h). Figure 5.7(c) shows the plate in a

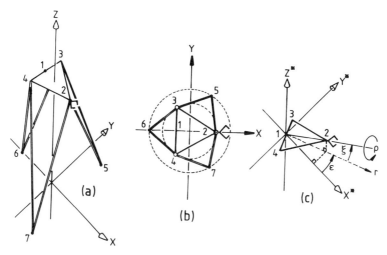

Figure 5.7 A parallel topology robot: (a) general arrangement (b) plan view of reference position (c) rotational coordinates.

disturbed position. The angles ρ, ξ and ε are the orientational degrees of freedom in world coordinates. With respect to a frame X^*Y^*Z, parallel to frame XYZ and with origin at point 1 on the plate, it can be shown (Appendix II) that

$$\left.\begin{aligned} x_2^* &= m \cos \xi \cos \varepsilon \\ y_2^* &= m \cos \xi \sin \varepsilon \\ z_2^* &= m \sin \xi \end{aligned}\right\} \tag{5.4}$$

where m is the distance between points 1 and 2 on the plate,

$$\left.\begin{aligned} x_3^* &= -n[\cos \rho \sin \varepsilon + \sin \rho \sin \xi \cos \varepsilon] \\ y_3^* &= n[\cos \rho \cos \varepsilon - \sin \rho \sin \xi \sin \varepsilon] \\ z_3^* &= n \sin \rho \cos \xi \end{aligned}\right\} \tag{5.5}$$

where n is half the length of the plate's side,

$$\left.\begin{aligned} x_4^* &= n[\cos \rho \sin \varepsilon + \sin \rho \sin \xi \cos \varepsilon] \\ y_4^* &= n[\sin \rho \sin \xi \sin \varepsilon - \cos \rho \cos \varepsilon] \\ z_4^* &= -n \sin \rho \cos \xi \end{aligned}\right\} \tag{5.6}$$

Knowing the coordinates of points 2, 3 and 4 in frame X^*Y^*Z it is possible to address the inverse problem of determining machine coordinates from world coordinates. If the world coordinate vector of the gripper is $[x_2, y_2, z_2, \xi, \varepsilon, \rho]^T$, what are the extensions of the various links from their reference positions? The procedure is as follows:

(1) Find $x_1y_1z_1$ from

$$\left.\begin{array}{l} x_1 = x_2 - x_2^* \\ y_1 = y_2 - y_2^* \\ z_1 = z_2 - z_2^* \end{array}\right\} \qquad (5.7)$$

(2) Find $x_3y_3z_3$ from

$$\left.\begin{array}{l} x_3 = x_1 + x_3^* \\ y_3 = y_1 + y_3^* \\ z_3 = z_1 + z_3^* \end{array}\right\} \qquad (5.8)$$

(3) Find $x_4y_4z_4$ from

$$\left.\begin{array}{l} x_4 = x_1 + x_4^* \\ y_4 = y_1 + y_4^* \\ z_4 = z_1 + z_4^* \end{array}\right\} \qquad (5.9)$$

Knowing the coordinates of the three plate fixing points in world coordinates, it is a straightforward task to determine the lengths of the six links from

$$L_{ij} = \sqrt{(x_j - x_i)^2 + (y_j - y_i)^2 + (z_j - z_i)^2} \qquad (5.10)$$

where $j(2, 3, 4)$ defines an endpoint on the plate and $i(5, 6, 7)$ a corresponding endpoint on the earth.

The change in a machine coordinate is determined from

$$\Delta L_{ij} = L_{ij} - L \qquad (5.11)$$

where L, the reference length, is given by

$$L = \sqrt{R(R - r) + r^2 + h^2}$$

An example will help to clarify the procedures. Take $R = 1.5$ m, $r = 1$ m, $h = 5$ m and assuming an application in which it is necessary to change the end-effector's world coordinate vector from an initial value of $[1, 0, 5, 0, 0, 0]^T$ to a final value of $[1, 1, 6, 30°, 30°, 30°]^T$. In addition, in completing this movement the end-effector must follow a straight-line path at a speed of 0.0707 m/s. The length of the path is calculated to be 1.414 m, so a time of 20 s will be required for the movement. Table 5.1 shows the required machine coordinates at 2 s intervals. In practice the number of increments would be much greater than this, allowing updates to servomotors to be made at intervals around 0.2 s. Note also that coordinated control is achieved, i.e. all machine coordinates reach their final values simultaneously.

Table 5.1

t	x_2	y_2	z_2	$\xi°$	$\rho°$	$\varepsilon°$	ΔL_{27}	ΔL_{25}	ΔL_{35}	ΔL_{36}	ΔL_{46}	ΔL_{47}
0	1	0	5	0	0	0	0	0	0	0	0	0
2	1	0.1	5.1	3	3	3	0.122	0.073	0.074	0.060	−0.017	−0.033
4	1	0.2	5.2	6	6	6	0.246	0.149	0.147	0.120	−0.030	−0.067
6	1	0.3	5.3	9	9	9	0.370	0.227	0.218	0.179	−0.040	−0.101
8	1	0.4	5.4	12	12	12	0.494	0.308	0.288	0.239	−0.044	−0.132
10	1	0.5	5.5	15	15	15	0.620	0.391	0.356	0.298	−0.044	−0.159
12	1	0.6	5.6	18	18	18	0.747	0.477	0.421	0.356	−0.037	−0.180
14	1	0.7	5.7	21	21	21	0.874	0.565	0.483	0.414	−0.022	−0.192
16	1	0.8	5.8	24	24	24	1.001	0.655	0.542	0.470	0.002	−0.195
18	1	0.9	5.9	27	27	27	1.123	0.747	0.597	0.526	0.035	−0.186
20	1	1	6	30	30	30	1.258	0.841	0.648	0.581	0.078	−0.160

Kinematic control

It has been demonstrated that geometric control, although exact, involves complex calculations which must be performed in a short time. Kinematic control simplifies the calculations by using the approximate method of small perturbations to give velocity control starting from a known position. The method was introduced in Chapter 2, where the Jacobian J of matrix T was defined as

$$J = \frac{\partial X}{\partial \Theta} = \frac{\partial T}{\partial \Theta} \qquad (5.12)$$

The kinematic method used the inverse of this equation,

$$\Delta \Theta = J^{-1} \Delta X \qquad (5.13)$$

to generate the incremental values of Θ required for the increments ΔX. The Jacobian is in general a non-linear function of Θ, and so the values of its elements change as the manipulator moves. Computer controls normally update position values at regular time intervals, and the following procedure generates a stream of incremental values $\Delta \Theta$ which define speeds of machine joints or can be used to update required machine coordinate values:

Step 1. calculate ΔX for required velocity
Step 2. formulate the Jacobian
Step 3. calculate J for the present Θ
Step 4. calculate J^{-1}
Step 5. calculate $\Delta \Theta$
Step 6. increment Θ by $\Delta \Theta$
Step 7. return to Step 3

Consider the polar configuration shown in Figure 5.8(a) which has a pitch actuator on the wrist. There are four degrees of freedom; L, θ_1, θ_2 and

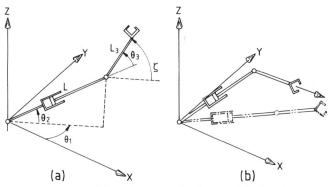

Figure 5.8 A four-degrees-of-freedom manipulator: (a) the degrees of freedom L, θ_1, θ_2 and θ_3: ξ is the pitch angle (b) manipulator at start and at destination.

θ_3. The tool centre point (TCP) is to be moved from the start coordinates

$$X_S = (x, y, z, \xi)^T = (1, 0, 1, 0)^T$$

to the destination coordinates

$$X_D = (1, 1, 0, 0)^T \qquad \text{(Figure 5.8(b))}$$

in 10 s, and the computer updates values every 100 ms. What machine coordinates are required at each time interval to achieve a straight-line motion at constant speed?

Using the shorthand notation introduced in Chapter 2, the relationship between world and machine coordinates is given by

$$
\begin{aligned}
x &= (LC2 + L_3 C23)C1 \\
y &= (LC2 + L_3 C23)S1 \\
z &= LS2 + L_3 S23 \\
\xi &= 23
\end{aligned}
\qquad (5.14)
$$

where $L_3 = 0.25$ m. The seven steps outlined above may now be applied to this example.

Step 1. The required increments of world coordinates are

$$\Delta x = (0, 0.01, -0.01, 0)^T.$$

Step 2. The Jacobian is constructed from eqns (5.14) by partially differentiating each expression with respect to each of the four machine coordinates:

$$
J = \begin{bmatrix}
-(LC2 + L_3 C23)S1 & -(LS2 + L_3 S23)C1 & C2C1 & -L_3 S23 C1 \\
LC2 + L_3 C23 & -(LS2 + L_3 S23)S1 & C2S1 & -L_3 S23 S1 \\
0 & LC2 + L_3 C23 & S1 & L_3 C23 \\
0 & 1 & 0 & 1
\end{bmatrix}
$$

Step 3. The machine coordinates corresponding to the start position are

$(0, 0.9273, 1.25, -0.9273)$ (Note that angles are expressed in *radians*.) The Jacobian at the start position is therefore

$$J = \begin{bmatrix} 0 & -1 & 0.6 & 0 \\ 1 & 0 & 0 & 0 \\ 0 & 1 & 0.8 & 0.25 \\ 0 & 1 & 0 & 1 \end{bmatrix}$$

Step 4. Inversion of the Jacobian yields

$$J^{-1} = \begin{bmatrix} 0 & 1 & 0 & 0 \\ -0.64 & 0 & 0.48 & -0.12 \\ 0.6 & 0 & 0.8 & -0.2 \\ 0.64 & 0 & -0.48 & 1.12 \end{bmatrix}$$

Step 5. Machine coordinate increments may now be calculated:

$$J^{-1}\Delta X = \begin{bmatrix} 0 & 1 & 0 & 0 \\ -0.64 & 0 & 0.48 & -0.12 \\ 0.6 & 0 & 0.8 & -0.2 \\ 0.64 & 0 & -0.48 & 1.12 \end{bmatrix}\begin{bmatrix} 0 \\ 0.01 \\ -0.01 \\ 0 \end{bmatrix} = \begin{bmatrix} +0.01 \\ -0.0048 \\ -0.008 \\ +0.0048 \end{bmatrix} = \Delta\Theta$$

Step 6. New values of Θ are obtained:

$$\Theta = \begin{bmatrix} 0 \\ 0.9273 \\ 1.25 \\ -0.9273 \end{bmatrix} + \begin{bmatrix} +0.01 \\ -0.0048 \\ -0.008 \\ +0.0048 \end{bmatrix} = \begin{bmatrix} 0.01 \\ 0.9225 \\ 1.242 \\ -0.9225 \end{bmatrix}$$

Step 7. Returning to Step 3, a new Jacobian is calculated from the values generated in Step 6:

$$J = \begin{bmatrix} -0.0100 & -1.0000 & 0.6038 & 0.0000 \\ 1.0000 & -0.0100 & 0.0060 & 0.0000 \\ 0.0000 & 1.0000 & 0.7971 & 0.2500 \\ 0.0000 & 1.0000 & 0.0000 & 1.0000 \end{bmatrix}$$

This is inverted and used to calculate the next machine coordinate increments.

The kinematic method may be evaluated in this example by comparing the machine coordinates it generates for the first few steps with the exact values, which the geometric method obtains by inverting eqn (5.14). The figures are given in Table 5.2

Table 5.2.

Kinematic solution	Geometric solution	
(increments)	(machine coordinates)	
Start	0.0000	0.0000
	0.9273	0.9273
	1.2500	1.2500
	−0.9273	−0.9273
+0.0100	0.0100	0.0100
−0.0048	0.9225	0.9225
−0.0080	1.2420	1.2420
+0.0048	−0.9225	−0.9225
+0.0100	0.0200	0.0200
−0.0048	0.9177	0.9174
−0.0079	1.2340	1.2342
+0.0049	−0.9176	−0.9174
+0.0100	0.0300	0.0300
−0.0051	0.9126	0.9124
−0.0078	1.2262	1.2264
+0.0051	−0.9125	−0.9124
+0.0100	0.0400	0.0400
−0.0052	0.9074	0.9071
−0.0077	1.2185	1.2187
+0.0052	−0.9073	−0.9071

5.5 Continuous control

The geometric and kinetic control techniques described earlier generate the required values and rates of change of machine coordinates. These signals are set-point inputs r to control systems for each axis, and it is to these systems that we now turn our attention.

In many cases it is necessary to alter the characteristics of a system, e.g. to improve stability. Two important compensation techniques will be described which will enable the transfer function G_s of a system to be moulded to requirements. Feedback compensation involves modification to the feedback signal by transfer function H (Figure 5.9(a)), and cascade compensation operates on the error signal by transfer function G_c (Figure 5.9(b)).

The practical value of the compensators is that they can alter the system characteristics in a desirable and predictable manner, avoiding changes in major mechanical or electrical features of an otherwise undesirable system. For the general case with both feedback and cascade compensation, the

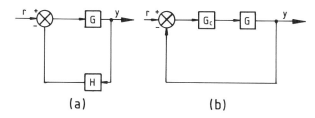

Figure 5.9 Closed-loop systems: (a) feedback compensation H (b) cascade compensation G_c.

system transfer function is

$$\frac{y}{r}(s) = \frac{G_s G_c}{1 + G_s G_c H} = \frac{G}{1 + GH}$$ (5.15)

It was seen earlier that robot dynamics are highly non-linear, making an exact analysis extremely difficult. However, at slow speeds and for small changes in coordinates the robot system behaves in a linear manner and a linear analysis can give useful results.

Linear systems

A linear system is defined as one whose behaviour may be described by a set of first-order linear ordinary differential equations; for a system with input u and output y this definition may be summarized in the state equations

$$\left.\begin{array}{l} \dot{x} = Ax + Bu \\ y = C^T x \end{array}\right\}$$ (5.16)

where x is the state vector, A is an $n \times n$ matrix, n is the number of the states and therefore the order of the system, and B and C are the input and output vectors. When the system is linear all the elements of A, B and C are scalar quantities.

Many single-input–single-output control systems approximate to a second-order system, with transfer function of the form

$$\frac{y}{r}(s) = \frac{\omega_n^2}{s^2 + 2\zeta\omega_n s + \omega_n^2}$$ (5.17)

where s is the Laplace operator, ζ is the damping ratio and ω_n is the natural frequency, which is the frequency of the response in the absence of damping.

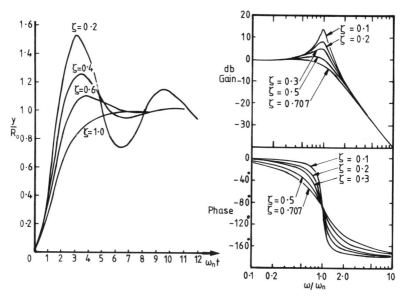

Figure 5.10   Responses of a second-order system: (a) to a step input R_0 (b) to an input of frequency ω.

The response of such a system to a step input R_0 is given by

$$\zeta < 1: \frac{y}{R_0} = 1 - \frac{1}{\sqrt{1-\zeta^2}}\exp\left(-\zeta\omega_n t\right)\sin\omega_n\sqrt{1-\zeta^2}t + \cos^{-1}\zeta$$

$$\zeta > 1: \frac{y}{R_0} = 1 - \frac{\zeta + \sqrt{\zeta^2-1}}{2\sqrt{\zeta^2-1}}\exp\left((-\zeta + \sqrt{\zeta^2-1})\omega_n t\right) \qquad (5.18)$$

$$+ \frac{\zeta - \sqrt{\zeta^2-1}}{2\sqrt{\zeta^2-1}}\exp\left((-\zeta - \sqrt{\zeta^2-1})\omega_n t\right)$$

These are illustrated in Figure 5.10(a). The curves indicate transient and steady-state elements. For $\zeta > 0$ the output ultimately settles to a steady state R_0. If $\zeta < 1$ there is an oscillatory transient with a damped frequency $\omega_d = \omega_n\sqrt{1-\zeta^2}$. For $\zeta > 1$ the transient is monotonic.

The response to a step input can give useful information in the time domain: the frequency response, which is the steady-state response to sinusoidal inputs of different frequencies, can also be helpful. Figure 5.10(b) is a Bode diagram of the frequency response of a second-order system, excited by a sinusoidal signal of amplitude A and frequency ω. The gain in dB is 20 log (output amplitude/A). The phase lag of the output signal is also shown.

Resonance occurs near ω_n and grows in magnitude as ζ decreases. A high resonance peak may be damaging to mechanical equipment, and must be avoided by providing sufficient damping or by ensuring that ω_n is well above the highest frequency of input likely to occur. For example it is necessary to ensure that the natural frequency of a control system does not coincide with the natural frequency of the robot link which it is controlling. It has been suggested (Paul, 1982) that the former should be at least half the latter.

Stability

If transient oscillations do not decay, a system is unstable. Instability can be predicted from the transfer function of the system. The denominator is extremely important since its form determines the characterization of the transient response: for this reason the equation formed by equating the denominator to 0 is known as the characteristic equation. For example the characteristic equation for the transfer function of eqn (5.17) is

$$s^2 + 2\zeta\omega_n s + \omega_n^2 = 0 \tag{5.19}$$

The roots of this equation are

$$\begin{aligned}\zeta > 1 && s = -\zeta\omega_n \pm \omega_n\sqrt{\zeta^2 - 1} \\ \zeta < 1 && s = -\zeta\omega_n \pm j\omega_n\sqrt{1 - \zeta^2}\end{aligned} \Bigg\} \tag{5.20}$$

The real parts of these roots determine the stability of the system. For example in eqns (5.18), the real roots can be identified with the exponential function. If the real roots are negative the exponential ultimately decays to 0—the system is stable: if positive the exponential grows with time and the system is unstable.

Thus a useful test for stability is to examine the roots of the characteristic equation. In the case of a closed-loop system this equation takes the form

$$1 + G(s)H(s) = 0 \tag{5.21}$$

These roots are also called the eigenvalues, and for stability they must have negative real parts.

The eigenvalues may be calculated directly from the state-space representation of a system, since the characteristic equation is

$$|sI - A| = 0 \tag{5.22}$$

where I is the unit matrix. For example, the state-space representation of the second-order system is

$$\begin{bmatrix} \dot{x}_1 \\ \dot{x}_2 \end{bmatrix} = \begin{bmatrix} 0 & 1 \\ -\omega_n^2 & -2\zeta\omega_n \end{bmatrix} \begin{bmatrix} x_1 \\ x_2 \end{bmatrix} + \begin{bmatrix} 0 \\ \omega_n^2 \end{bmatrix} r \tag{5.23}$$

$$y = [1 \quad 0][x_1 \quad x_2]^\mathrm{T} \tag{5.24}$$

which gives the characteristic equation

$$|sI - A| = \begin{vmatrix} s & -1 \\ \omega_n^2 & (s + 2\zeta\omega_n) \end{vmatrix} = s^2 + 2\zeta\omega_n s + \omega_n^2 = 0 \tag{5.25}$$

Single-axis control

The geometric control method previously explained relied upon the existence of a closed-loop control system to control the position of each joint of
the robot. The control techniques discussed in Section 5.5 may now be
developed to provide this position control. Figure 5.11 shows an inertial
load, with viscous friction, being driven by an electric motor. It is assumed
that the torque from the motor is proportional to its voltage input. The
parameters are

amplifier input	e (V)
amplifier gain	A (V/V)
amplifier output	e_a (V)
motor torque constant	M (Nm/V)
motor torque	T (Nm)
load inertia	J (Nm^2)
viscous friction constant	B (Nm/rad/s)
load position	θ (rad)
potentiometer constant	A_p (volts/rad)
tachometer constant	A_t (volts/rad/s)

Applying Newton's second law to the system gives

$$T = J\ddot{\theta} + B\dot{\theta} \tag{5.26}$$

giving a transfer function

$$\frac{\theta}{T}(s) = \frac{1}{Js^2 + Bs} \tag{5.27}$$

A position-control system using only the potentiometer signal may be

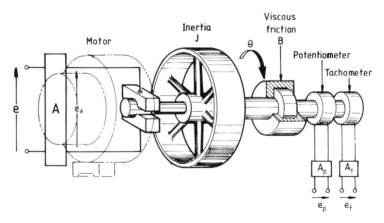

Figure 5.11 Electric motor driving load with viscous friction, incorporating position
and velocity measurement.

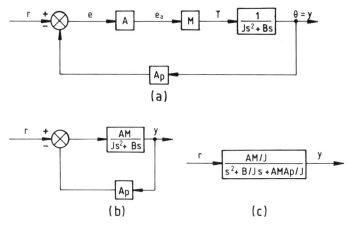

Figure 5.12 Position control with position feedback: (a) control of system of Figure 5.11 (b) forward path and feedback transfer functions (c) closed-loop transfer function.

constructed as shown in Figure 5.12(a), where r is the set-point and θ is the output y. This control system reduces (Figure 5.12(b) and (c)) to a second-order system with

$$\omega_n = \sqrt{\frac{AMA_p}{J}} \quad \text{and} \quad \zeta = \frac{B}{2\sqrt{JAMA_p}} \tag{5.28}$$

and a steady-state gain $K = 1/A_p$.

Feedback compensation

It is often necessary to alter the performance of a control system, and eqns (5.28) show that although this may be done by altering the amplifier gain A, it is not possible to have independent control of ω_n and ζ by this means.

A common requirement is to increase ζ to reduce or eliminate the maximum overshoot in the transient response. Velocity feedback is a method of increasing ζ by adding compensation to the feedback loop. Since a tachometer generates a voltage e_t proportional to velocity, its transfer function is

$$\frac{e_t}{\theta}(s) = A_t s \tag{5.29}$$

The voltage e_t from the tachometer (Figure 5.11) is added to the potentiometer voltage to form a compensated feedback signal (Figure 5.13(a)). Reducing the system to the closed-loop transfer function (Figures 5.13(b) and (c)), shows that only the coefficient of s, and therefore the value of ζ, is changed from that in Figure 5.12(c).

This argument may now be extended to a geared electric motor drive typical of many robots (Figure 5.14). The motor has a resistance R_m and

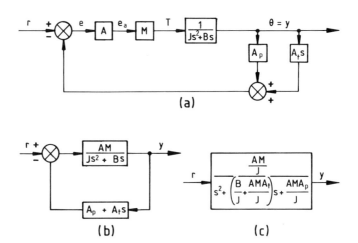

Figure 5.13 Position control with position and velocity feedback: (a) control of system of Figure 5.11 (b) forward path and feedback transfer functions (c) closed-loop transfer function.

inductance L_m and generates a back e.m.f. e_m which is proportional to its angular velocity ω_m. With a fixed field, the torque T from the motor is proportional to the current i. This current is measured by amplifying the voltage across a resistance R_i, where $R_i \ll R$. The load is driven to a position θ at velocity ω through a reduction gear of ratio $n:1$. The system is described by the equations

$$\left. \begin{array}{lll} e_a = Ae & e_m = C_B \omega_m & e_a - e_m = Ri + Li \\ \omega_m = n\omega & \omega = \dot{\theta} & T = C_T i = (J\omega_m + B\omega_m)/n^2 \end{array} \right\} \quad (5.30)$$

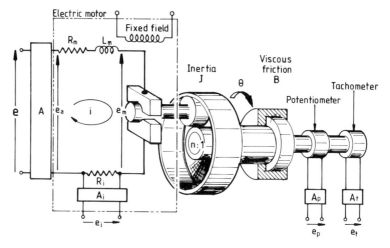

Figure 5.14 Armature-controlled electric motor driving load with viscous friction, incorporating position, velocity and armature current measurement.

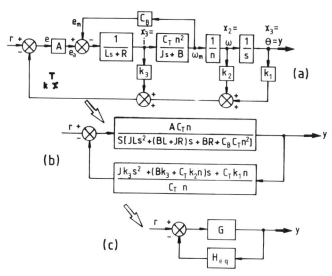

Figure 5.15 Position control with full state feedback: (a) control of system in Figure 5.14 (b) forward path and feedback transfer functions (c) H equivalent reduction.

and the resultant transfer functions are

$$\left.\begin{array}{ccc} \dfrac{e_a}{e}(s) = A & \dfrac{e_m}{\omega_m}(s) = C_B & \dfrac{i(s)}{e_a - e_m} = \dfrac{1}{Ls + R} \\[3mm] \dfrac{\omega_m}{i}(s) = \dfrac{C_T m^2}{Js + B} & \dfrac{\omega}{\omega_m}(s) = \dfrac{1}{n} & \dfrac{\theta}{\omega}(s) = \dfrac{1}{s} \end{array}\right\} \quad (5.31)$$

The block diagram representation of these transfer functions (Figure 5.15(a)) indicates that the system is of third order, and since θ, ω and i are successively related by one order of integration, a valid state vector is

$$x = \begin{bmatrix} \theta \\ \omega \\ i \end{bmatrix} \quad (5.32)$$

State feedback is a method of achieving full control of the coefficients of the closed-loop transfer function, and is implemented by adding all amplified state signals to form the feedback signal

$$f = k^T x \quad (5.33)$$

where k is the feedback vector. In this system, for example,

$$f = \begin{bmatrix} k_1 & k_2 & k_3 \end{bmatrix} \begin{bmatrix} x_1 \\ x_2 \\ x_3 \end{bmatrix} = \begin{bmatrix} A_p & A_t & A_i \end{bmatrix} \begin{bmatrix} \theta \\ \omega \\ i \end{bmatrix} = A_p\theta + A_t\omega + A_i i \quad (5.34)$$

The feedback coefficients k_1, k_2 and k_3 are in fact the gain settings of the potentiometer, tachometer and current sensor amplifiers respectively. To analyse the effect of state feedback on the closed-loop system, the block diagram can be reduced to the familiar form of closed-loop system (Figures 5.15(a)–(c)). This process is called the *H*-equivalent reduction, and generates a closed-loop transfer function

$$\frac{y}{r}(s) = \frac{G(s)}{1 + G(s)H_{eq}(s)}$$

$$= \frac{AC_T n}{JLs^3 + (BL + JR + AJk_3)s^2 + (BR + C_B C_T n^2 + ABk_3 \atop + AC_t k_2 n)s + AC_T k_1 n} \qquad (5.35)$$

All the coefficients of the closed-loop transfer function may be set to desired values by adjusting A, k_1, k_2 and k_3.

Stability of the system may again be determined by examining the three roots of the third-order characteristic equation.

When the equations $\dot{x} = Ax + Bu$ and $y = C^T x$ represent an open-loop system and full state feedback is to be used, the closed-loop transfer function is

$$\frac{y}{r}(s) = C^T(sI - A + Bk^T)^{-1}B \qquad (5.36)$$

and the characteristic equation is

$$|sI - A + Bk^T| = 0 \qquad (5.37)$$

where k is the feedback vector. The reader may wish to confirm that the state equations

$$\begin{bmatrix} \dot{x}_1 \\ \dot{x}_2 \\ \dot{x}_3 \end{bmatrix} = \begin{bmatrix} 0 & 1 & 0 \\ 0 & (-B/J) & (C_T n/J) \\ 0 & (-C_B n/L) & (-R/L) \end{bmatrix} \begin{bmatrix} x_1 \\ x_2 \\ x_3 \end{bmatrix} + \begin{bmatrix} 0 \\ 0 \\ A/L \end{bmatrix} u \qquad (5.38)$$

$$y = \begin{bmatrix} 1 & 0 & 0 \end{bmatrix} x$$

are a valid representation for the example given, and that eqns (5.36) to (5.38) yield a closed-loop transfer function in agreement with eqns (5.35).

Cascade compensation

We have seen that the addition of a velocity (i.e. derivative) signal to the position (i.e. proportional) feedback has a stabilizing effect on a control system. The further addition of an integral signal can eliminate some steady-state errors. The most common type of cascade compensator is the proportional-integral-derivative (PID) or three-term controller. The PID controller operates on the error signal, and has a transfer function of the

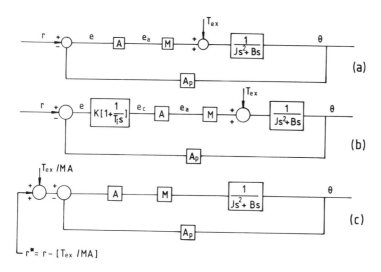

Figure 5.16 Eliminating steady-state errors due to gravity loads: (a) closed loop with disturbance torque T_{ex} (b) proportional plus integral control (c) amended set-point.

form

$$G_c(s) = K\left(1 + \frac{1}{T_i s} + T_d s\right) \qquad (5.39)$$

where K is the gain, T_i is the integral time constant and T_d is the derivative time constant.

The effect of the integral term is demonstrated by reference to the position control system of Figure 5.11. The block diagram in Figure 5.16(a) is similar to that of Figure 5.12(a), but on this occasion a steady disturbance torque T_{ex} has been added to the system. It was shown earlier that gravity components provide such torques in practical systems. The torque applied to the load is now

$$T = Me_a + T_{ex} \qquad (5.40)$$

and the output can be determined from the equation

$$(Js^2 + Bs + MAA_p)\theta = MAr + T_{ex} \qquad (5.41)$$

The steady-state condition is determined by setting $s = 0$ (rates of change are 0), giving

$$\theta_{ss} = (r_{ss}/A_p) + (T_{ex}/MAA_p) \qquad (5.42)$$

Thus the presence of the steady disturbance torque has given rise to an output error of magnitude (T_{ex}/MAA_p).

This steady-state error can be eliminated by use of an integral term in the controller equation. Neglecting the derivative term in the PID controller, we have

$$\frac{e_c}{e}(s) = K\left(1 + \frac{1}{T_i s}\right) \tag{5.43}$$

where e_c is a compensated signal.

The resulting block diagram is shown in Figure 5.16(b). It is easy to show that

$$\left[Js^2 + Bs + MAKA_p\left(1 + \frac{1}{T_i s}\right)\right]\theta = MAK\left(1 + \frac{1}{T_i s}\right)r + T_{ex}$$

or   (5.44)

$$[Js^3 + Bs^2 + MAKA_p s + MAKA_p/T_i]\theta = [MAKs + MAK/T_i]r + sT_{ex}.$$

The characteristic equation is now of third order. Steady-state conditions can be determined by setting $s = 0$, giving

$$\theta_{ss} = r_{ss}/A_p \tag{5.45}$$

which is the same as the original undisturbed and uncompensated system.

Other methods of eliminating errors due to gravity loads

If the disturbance torques are known (i.e. they are not random), then it is possible to compensate for their presence by amending the set-point. Figure 5.16(c) shows the required block-diagram manipulation. The resultant amended set-point is

$$r^* = r - (T_{ex}/MA) \tag{5.46}$$

These external torques can often be quite large in robot systems. They are also non-linear with respect to machine coordinates, and require a computer for their calculation. For example, in the three-degrees-of-freedom system of Figure 5.6 (partly repeated in Figure 5.17(a)) the gravity torques are

$$\text{Joint 2:} \quad T_{2ex} = -[M_a L_2 C\theta_2 + 0.5M_b L_3 C(\theta_2 + \theta_3)]g \tag{5.47}$$

$$\text{Joint 3:} \quad T_{3ex} = -[0.5M_b L_3 C(\theta_2 + \theta_3)]g \tag{5.48}$$

where $M_a = 0.5m_2 + m_{23} + m_3 + m_{34}$ and $M_b = m_3 + 2m_{34}$.

It is also possible to eliminate these disturbing torques by mechanical means (Figure 5.17(b)). This method is particularly useful for machines which are taught by lead-through, when it is desirable that a robot maintains a given position when the manual force is removed. Figure 5.17(b) shows a mass M_4 supported by a light parallel linkage. Taking moments about elbow and shoulder (or using the Lagrange method of Appendix II) shows that the weight $M_4 g$ provides balancing torques T_{2b} and T_{3b} where

$$T_{2b} = M_4 g[L_4 C(\theta_2 + \theta_3) + L_5 C\theta_2] \tag{5.49}$$

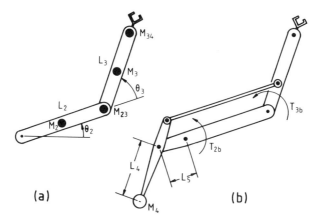

Figure 5.17 Balancing: (a) unbalanced manipulator (b) balanced manipulator.

and

$$T_{3b} = M_4 g L_4 C(\theta_2 + \theta_3) \tag{5.50}$$

Joint 3 is therefore balanced when

$$T_{3ex} + T_{3b} = 0$$

or when

$$M_4 L_4 = 0.5 M_b L_3 \tag{5.51}$$

Joint 2 is balanced when

$$T_{2ex} + T_{2b} = 0$$

and when eqn (5.51) is satisfied this reduces to

$$M_4 L_5 = M_a L_2 \tag{5.52}$$

Hence if M_4, L_4 or L_5 is arbitrarily set, eqns (5.51) and (5.52) allow the remaining two parameters to be determined.

5.6 Digital control

In order to drive the position control systems of a robot the computer must calculate desired machine coordinates, and in turn the necessary set-point values. The set-points demand that the control systems drive to the desired machine coordinates. The set-points are digital values, and are communicated to the control systems via digital-to-analogue converters (DAC) as shown in Figure 5.18(a).

The response of a continuous robot control system is not analytically predictable because of the non-linear nature of the robot dynamics, and to improve the control system performance over the whole range of movement

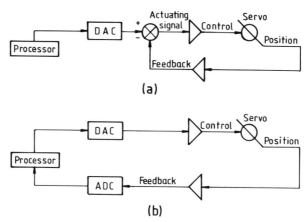

Figure 5.18 Position control of digital set-point: (a) conversion of set-point to analogue for continuous control system (b) control strategy implemented by processor algorithm.

it may be desirable to alter not just the set-point but also the parameters of the controller itself. Although this is possible with some continuous systems, a more practical and versatile solution is to make the processor part of the control system (Figure 5.18(b)) (see Baruh *et al.*, 1980; Turner *et al.*, 1984; Patzelt, 1982). The addition of an analogue-to-digital converter (ADC) is necessary to allow the processor to sample the feedback signal. The operation of ADCs and DACs is considered in Chapter 6.

Digital control system performance

The control system of Figure 5.18(a) is analysed by considering it to function as in Figure 5.19(a). The actuating signal is sampled at intervals of time *T*, the asterisk denoting a sampled quantity. The sampled actuating signal e^* is acted on by the digital controller, whose output m^*, converted to an analogue signal, derives the servomechanism. A common device for this DAC is the zero-order hold, whose input and output and Laplace equivalent are shown in Figures 5.19(b), (c) and (d).

Digital PID controller

The development of the control algorithm to simulate a PID controller will illustrate the principle (Erdelyi *et al.*, 1980). The derivative of a continuous function of time $f(t)$ is approximated by

$$\dot{f}(t) = \frac{f(k) - f(k-1)}{T} \tag{5.53}$$

where $f(k)$ is the present value of $f(t)$, $f(k-1)$ is the previous value of $f(t)$, i.e. $f(t-T)$, and T is the sampling interval.

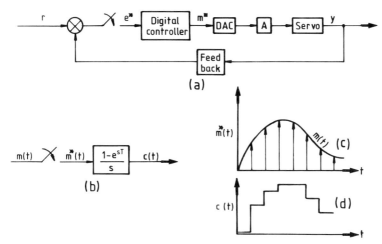

Figure 5.19 Digital control: (a) digital controller in continuous system (b) sampling with zero-order hold (c) waveform of controller output (d) zero-order-hold output.

The algorithm implements PID control:

$$m(s) = K\left(1 + \frac{1}{T_i s} + T_d s\right)e(s) \tag{5.54}$$

Rewriting this as

$$sm(s) = K\left[se(s) + \frac{1}{T_i}e(s) + T_d s^2 e(s)\right]$$

and noting that $sf(s)$ is the transform of $\dot{f}(t)$, we obtain

$$\frac{m(k) - m(k-1)}{T}$$

$$= K\left[\frac{e(k) - e(k-1)}{T} + \frac{1}{T_i}e(k) + T_d \frac{e(k) - 2e(k-1) + e(k-2)}{T^2}\right]$$

which generates the difference equation

$$m(k) = m(k-1) + K\left(1 + \frac{T}{T_i} + \frac{T_d}{T}\right)e(k) - k\left(1 + \frac{2T_d}{T}\right)e(k-1) + KT_d e(k-2)$$

$$= m(k-1) + a_1 e(k) - a_2 e(k-1) + a_3(k-2) \tag{5.55}$$

The difference equation for computing the error e is

$$e(k) = r(k) - f(k) \tag{5.56}$$

where $r(k)$ is the present set-point and $f(k)$ is the present feedback value. Equations (5.55) and (5.56) represent the calculation of the value of the controlling signal m^*, which is carried out at each sampling interval. It is necessary for the processor to store the previous value of its output, i.e.

$m(k-1)$, and the two previous values of the error, i.e. $e(k-1)$ and $e(k-2)$.

The formulation of the difference equation is based on eqn (5.53), which is an approximation. The sampling frequency is limited by processing speed, and as the sampling interval is shortened the performance of the digital controller more closely resembles that of its continuous equivalent.

5.7 Robot control systems

The above analyses illustrate the difficult nature of the robot control problem and re-emphasize the importance of the computer in robot control systems. Computational burdens are high even in those systems where the non-linear dynamics can be neglected and a kinematic analysis is satisfactory.

Position controllers

In position controllers the desired machine coordinates, determined by inversion of the world coordinates, are used as set-points to position servos on each axis. Earlier sections of this chapter and others have indicated the complexity of the inversion transformation. For example, the Unimate, in spite of kinematic features which ease the analysis, still requires the following operations: 30 multiply/divide, 15 add/subtract, 3 square root, 7 square, 6 cosine/sine and 6 arctangent. The use of parallel topology robots can reduce the amount of calculation, and in serial robots the equation can be greatly simplified if three intersecting axes are used, usually at the wrist.

Computation time presents a problem since it is usually necessary to send commands to the joint servos at around 50 Hz. If this is not possible then the method of trajectory planning may be used, in which the transformations are carried out and stored ahead of time. If operation in real time is essential then the use of rate control can reduce the computational load somewhat.

Rate controllers

In rate controllers, the inverse Jacobian is used to determine the set-points for rate servos.

We have

$$\dot{\Theta} = J^{-1}(\Theta)\dot{X} \qquad (5.57)$$

It would clearly be desirable to decouple the system so that a single input would control a single output. This can be achieved by making the world coordinate velocity vector $\dot{X}$ proportional to the world coordinate error vector $(X_i - X)$ or

$$\dot{X} = K(X_i - X) \qquad (5.58)$$

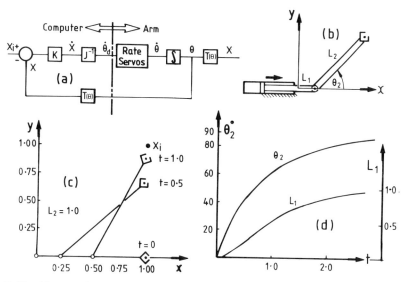

Figure 5.20 Decoupled rate control: (a) block diagram (b) a simple system with two degrees of freedom (c) configuration at various times (d) variation of machine coordinates with time.

where the gain matrix K has to be diagonal if decoupling is desired. Hence

$$\dot{\Theta} = J^{-1}(\Theta)K(X_i - X) \tag{5.59}$$

Figure 5.20(a) shows that the results of this calculation are used as set-points to rate servos. In order to determine the world coordinate vector X it is necessary to employ the forward transformation $X = T(\Theta)$.

A simple example will help to clarify the operation of such a system. Figure 5.20(b) shows a manipulator with two degrees of freedom, with one linear (L_1) and one rotary (θ_2) machine coordinate. We have

$$\left.\begin{array}{l} x = L_1 + L_2 C\theta_2 \\ y = L_2 S\theta_2 \end{array}\right\} \quad \text{or} \quad X = T(\Theta) \tag{5.60}$$

By differentiation

$$\begin{bmatrix} \dot{x} \\ \dot{y} \end{bmatrix} = \begin{bmatrix} 1 & -L_2 S\theta_2 \\ 0 & LC\theta_2 \end{bmatrix} \begin{bmatrix} \dot{L}_1 \\ \dot{\theta}_2 \end{bmatrix} \quad \text{or} \quad \dot{X} = J(\Theta)\dot{\Theta}$$

Hence

$$J(\Theta)^{-1} = \frac{1}{L_2 C\theta_2} \begin{bmatrix} L_2 C\theta_2 & L_2 S\theta_2 \\ 0 & 1 \end{bmatrix}$$

Figures 5.20(c) and (d) show the responses to a demand $[1, 1]^T$, starting from an initial world coordinate vector $[1, 0]^T$. Length L_2 is unity and an updating time of 0.1 s is assumed. The servo gains are set at 4 m/s/m. The following algorithm is employed:

Step 1. Calculate the initial X from the initial Θ i.e. use T
Step 2. Calculate $\dot{X}$ from $\dot{X} = K(X_i - X)$
Step 3. Calculate $\dot{\Theta}_d$ from $\dot{\Theta}_d = J^{-1}(\Theta)\dot{X}$
Step 4. Assume $\dot{\Theta} = \dot{\Theta}_d$
Step 5. Take $\Delta\Theta = 0.5\dot{\Theta}\,\Delta t$
Step 6. Update Θ by $\Delta\Theta$
Step 7. Calculate X from $X = T(\Theta)$
Step 8. Go to Step 2.

In practice it would be sufficient to update the value $J^{-1}(\Theta)$ every other sampling period. The problem of computing $T(\Theta)$ could be simplified if the measuring transducers incorporated sine and cosine encoders.

Dynamic control

A detailed study of control systems which takes into account the full dynamics of a robot is beyond the scope of this book. Earlier in this chapter the dynamic equations of a robot with three degrees of freedom were formulated. It was shown that, knowing the machine coordinates and their first and second derivatives, it was possible to calculate the forces and torques required of the various actuators. This forms the basis of a force/torque control system in which set-points to force/torque controllers are determined from the dynamic equations. However, a control system built on this basis would be open-loop in essence—it would be assumed that the applied torque would give the desired motions.

Decoupled closed-loop control involves a diagonal matrix relating $\ddot{X}$ and $(X_i - X)$: i.e.

$$\ddot{X} = K(X_i - X) \tag{5.61}$$

where X is the world coordinate vector. A block diagram of the resultant system is shown in Figure 5.21. There are three heavy computing demands— the transform from $\ddot{X}$ to $\ddot{\Theta}$, the determination of torques, and the forward transform T from machine coordinates to world coordinates. The last two of

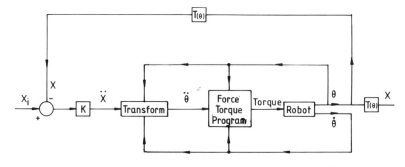

Figure 5.21   An elementary form of decoupled acceleration control.

these have already been discussed. The first needs more clarification: in general

$$\dot{X} = J(\Theta)\dot{\Theta}$$

hence

$$\ddot{X} = J(\Theta)\ddot{\Theta} + L(\Theta, \dot{\Theta}) \tag{5.62}$$

For example, in the earlier illustration of a rate-controlled system, this relationship took the form

$$\left.\begin{aligned}\ddot{x} &= \ddot{L}_1 - \ddot{\theta}_2 L_2 S\theta_2 - \dot{\theta}_2^2 L_2 C\theta_2 \\ \ddot{y} &= \ddot{\theta}_2 L_2 C\theta_2 - \dot{\theta}_2^2 L_2 S\theta_2\end{aligned}\right\} \tag{5.63}$$

Thus, knowing the $\ddot{x}$ and $\ddot{y}$ signals from the controller K, and the sampled machine coordinates and their velocities, it is possible to solve for the accelerations $\ddot{L}_1$ and $\ddot{\theta}_2$. These in turn are input to the force/torque program, which also needs the sampled values of Θ and $\dot{\Theta}$. The overall objective is to obtain a decoupled control system in which one input controls one world coordinate.

The system of Figure 5.21 would tend to give an oscillatory response and some form of stability augmentation would be required in practice: perhaps feedback of the rate of change of the world coordinate vector. A more detailed account of such controllers can be found in Lee (1982), Luh *et al.* (1980b) and Coiffet (1983a).

5.8   Recapitulation

The PLC is a suitable sequence controller for pick-and-place devices, where position control is by mechanical stops. Closed-loop position control is required for a robot, and the dynamic model shows it to be highly non-linear. Digital systems allow parameters to be altered to give optimum control throughout the working volume, but continuous control systems are still widely used. A computer is necessary to perform the complex transformations between world coordinates and machine coordinates. Geometric and kinematic control are two solutions which rely solely on the kinematics of the manipulator and do not take dynamic effects into account. Use of a rate controller allows decoupled position control of a robot, and full dynamic control is possible although heavy computing demands must be met.

Chapter 6

Feedback components

6.1 Introduction

It has been demonstrated that closed-loop control is required if a robot arm is to be accurately positioned in the presence of disturbances such as gravity loads. Since computation is necessary to transform from world coordinates, the required values of machine coordinates originate as digital information from a computer. Figure 6.1 shows four designs of control systems, each of which could set the elbow of a robot at an angle θ, the set-point value of θ being the digital output of a processor. In all parts of Figure 6.1, a solid line represents an analogue signal, and a broken line a digital signal.

Figure 6.1(a) shows the set-point transformed to an analogue quantity by a digital-to-analogue converter (DAC), and this signal drives a conventional continuous control system. Position and velocity feedback signals are subtracted from the input, and the resultant signal is operated on by a controller, amplified, and drives an electrohydraulic servovalve and actuator. If the position-measuring device was in this case a resolver, the position signal would be represented by the relative phase shift of two sine waves. This needs to be transformed into a voltage compatible with the velocity feedback signal, which might be a DC voltage from a tachometer. The amplifier accepts a voltage signal from the controller, and generates a current output to drive the electrohydraulic servovalve.

Figure 6.1(b) shows one method of implementing digital control of the same system. The analogue position signal is fed to an analogue-to-digital converter (ADC) which supplies the processor with a digital representation of position. The control strategy is implemented by the processor; no feedback of velocity is needed since it may be inferred from the rate of

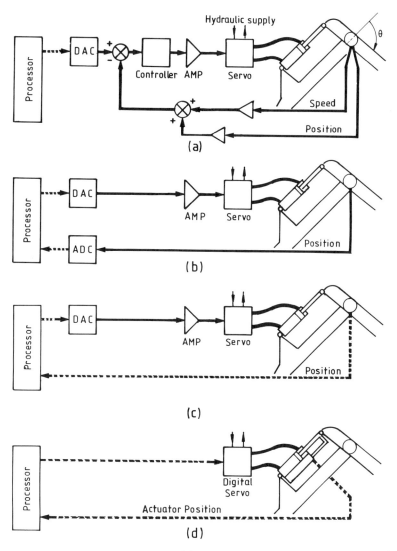

Figure 6.1 Position-control system: (a) continuous closed-loop control (b) digital control with analogue measurement (c) digital control with digital measurement (d) digital control with digital servo.

change of position by the control algorithm. Note that the output of the processor here is related to the error, and not to the set-point as in Figure 6.1(a).

The use of a digital position-measuring device, i.e. an encoder, eliminates the need for an ADC, simplifying the system to that of Figure 6.1(c). One concept of the future direction of robots (Albus, 1981) is shown in Figure 6.1(d). The DAC has been eliminated by using a digital servo, a device which directly accepts digital signals and generates a proportional flow rate of hydraulic fluid. Note that here the encoder measures the position of the cylinder rod rather than of the joint.

The foregoing discussion highlights some of the sensing and measurement hardware and techniques which are necessary for the implementation of closed-loop control, whether the controlled variable is position, speed or in some cases, force. The topics to be covered in this chapter are, therefore:

- position sensing and measurement
- velocity measurement
- force measurement
- digital-to-analogue conversion
- analogue-to-digital conversion
- interfacing

Although some of the material in this chapter may be applicable to the external sensors reviewed in Chapter 8, we are here concerned primarily with sensing and measurement internal to the robot, i.e. the methods employed to enable direct position or path control of the end-effector.

6.2   Position sensing and measurement

There are many devices for measuring position, and they give a variety of signals which may or may not be directly compatible with a particular control system. This section is concerned with the hardware for position measurement; the necessary processing of the signals for compatibility is covered in Sections 6.5 and 6.6.

It is important to note that position sensing, the detection of the fact that a certain position has been achieved, is distinct from position measurement, which involves a transducer attached to the point of interest from whose output a measurement of position can be inferred to some degree of accuracy.

All of the devices described in this section include a moving part which is attached to one link, the body of the device being held by the link on the other side of the joint. Thus they all measure the relative motion of two links of a manipulator i.e. a machine coordinate, although the geometry of the motion of the joint need not be the same as that of the measuring device.

Position measurement by variable resistance

A potentiometer may be used to measure linear or rotary position. Figures 6.2(a) and (b) show respectively linear and rotary potentiometers. Movement of the input shaft causes the resistance of the device to vary from 0 to maximum, the circuit being shown in Figure 6.2(c).

The resistance in the stator of a potentiometer may be wire-wound or constructed of a conducting plastic. In reality, therefore, the sliding contact

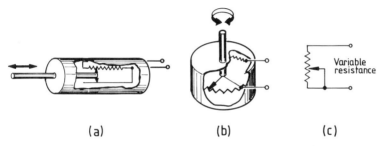

(a)   (b)   (c)

Figure 6.2 Potentiometer: (a) linear (b) rotary (c) electrical circuit.

of a wire-wound potentiometer makes contact only at n points along the travel, where n is the number of turns in the resistance. This affects the resolution of the device; for example the resolution of a 100-turn potentiometer cannot be better than 1%. Conductive plastic potentiometers, although not suffering from this phenomenon, are more susceptible to temperature effects.

Position measurement by variable differential transformer

The linear variable differential transformer (LVDT) consists of two identical secondary windings excited by one primary winding. The secondary windings are connected in opposition, as shown in Figure 6.3(a), so that the output is effectively the difference in their voltages. With a ferromagnetic core in the central (null) position, or removed completely, the mutual inductance between the primary winding and each of the secondary windings is equal; the voltage across each secondary winding is therefore equal and the differential output voltage is 0. The rotary variable differential transformer (RVDT) produces a similar effect over 180° by the rotation of a specially shaped core (Figure 6.3(b)). Within limits the voltage of the differential

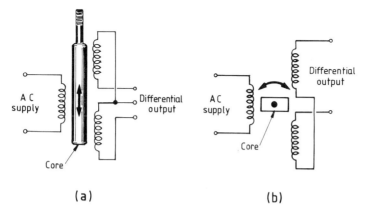

(a)   (b)

Figure 6.3 Variable differential transformer: (a) linear (LVDT) (b) rotary (RVDT).

output is proportional to the displacement of the core from the null position, and the phase shift of the output indicates the direction of the displacement.

Position measurement by resolver

Figure 6.4(a) shows a simplified view of the construction of a resolver, a device which can give a very accurate measurement of angular position. The stator has two windings, disposed at 90° to each other, to which are applied AC voltages with a 90° phase difference (Figure 6.4(b)). The output voltage is collected via brushes and slip rings from the rotor winding; it is an AC voltage of the same amplitude as the inputs, but shifted in phase.

Consider the rotor sitting at an angle θ, defined in Figure 6.4(a); it will pick up a component $V \cos \theta \cos \omega t$ from the cosine input, and a component $V \sin \theta \sin \omega t$ from the sine input.

The output voltage V_0 is the sum of these components:

$$\left. \begin{aligned} V_0 &= V \cos \theta \cos \omega t + V \sin \theta \sin \omega t \\ &= V \cos (\theta - \omega t) \end{aligned} \right\} \qquad (6.1)$$

This result may be confirmed by considering Figure 6.4(b). The resultant magnetic field due to the cosine and sine inputs has a constant amplitude, and rotates at a uniform angular velocity ω rad/s. At any instant the angle between the rotor and the resultant field is $(\theta - \omega t)$, so that the voltage induced across the rotor is V multiplied by the cosine of the angle between V and the rotor, confirming eqn (6.1).

In Figure 6.4(a) and (b) the rotor is shown at an angle θ of 60° to the reference, which is the axis of the cosine input winding. Figure 6.4(c) shows the relationship of the $V \cos (\theta - \omega t)$ output signal to the input signals for $\theta = 60°$. Since the quantity to be measured is represented by the relative phase shift of AC signals, the resolver is therefore relatively free of errors arising from voltage drops or electrical noise induced in the lines from the oscillator and measurement circuitry.

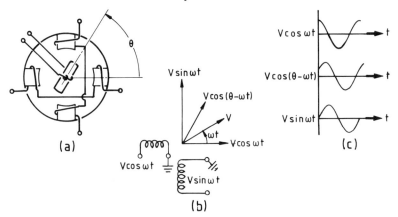

Figure 6.4 Resolver: (a) construction (b) applied and resolved voltages (c) phase relationships.

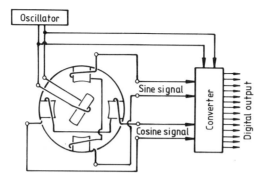

Figure 6.5 Digital absolute measuring system using brushless resolver.

Brushless resolvers are available in which the output signal is generated by a rotating transformer, eliminating the need for brushes and slip rings. Figure 6.5 illustrates a practical example of the use of a brushless resolver (Lanton, 1981). In this system an AC signal is supplied by an oscillator to the rotor of the resolver, and the sine and cosine components induced in the stator windings are monitored. All signal processing and computation circuitry is contained in one large-scale-integrated (LSI) package, which delivers a 14-bit digital version of the measured angle, updated at regular intervals. The angle of the rotor could be determined from the amplitude of either the cosine or sine signals, but error due to voltage drop is minimized by calculating the angle from the ratio of the signals.

Absolute encoders

A microprocessor-based control system requires measurement information to be presented in digital form. Techniques for transforming analogue measurements to digital information are covered later; here we consider absolute encoders, devices which output position information directly in digital form.

Figure 6.6(a) shows the essential elements of a rotary encoder. A narrow sheet of light, obtained by masking the illumination from a source with a slit, is directed at an array of photocells. The sheet of light is interrupted by a rotating disc, and the light patterns received by each photocell at its particular radial location are determined by the arrangement of opaque and transparent elements around the corresponding ring on the disc. Each photocell is turned ON (a digital 1) if it receives light through a transparent part of the disc, and OFF (a digital 0) if the light is blocked by an opaque part of the disc.

Figure 6.6(b) shows the disc pattern for a four-bit encoder. In a given signal the outer ring contains the least significant bit. The pattern of each number on the disc is dictated by the Gray code—not by the binary number system. The advantage of the Gray code over binary numbers for encoders is seen by comparing the binary and Gray code representations of the sixteen positions of a four-bit encoder (Table 6.1).

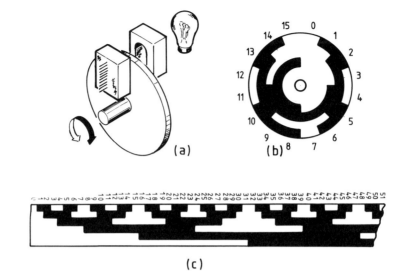

(a)   (b)

(c)

Figure 6.6 Absolute position encoder: (a) construction (b) four-bit rotary disc (c) part of six-bit linear mask.

Table 6.1

Gray code	Decimal position	Binary number
0000	0	0000
0001	1	0001
0011	2	0010
0010	3	0011
0110	4	0100
0111	5	0101
0101	6	0110
0100	7	0111
1100	8	1000
1101	9	1001
1111	10	1010
1110	11	1011
1010	12	1100
1011	13	1101
1001	14	1110
1000	15	1111
0000	0	0000

Anomalies can arise at the changeover from one segment to the next. Consider the disc to be moving from position 7 to 8. With respect to the binary numbers, if the change of status of the most significant bit from 0 to 1 is detected before the other three bits go to 0, there will be a position between 7 and 8 where the device will output 1111: this would be interpreted as decimal position 15. These false intermediate readings are eliminated by the Gray code, since at any point on the scale only one bit changes at a time. Gray code can be extended to any number of bits while retaining this characteristic and, similar to binary numbers, n bits are sufficient to define 2^n positions. Figure 6.6(c), for example, shows part of a six-bit linear encoder which will count from 0 to 63.

Incremental encoders

Figure 6.7(a) shows an incremental encoder which, regardless of resolution, requires only three photocells and two or three patterned rings to achieve position measurement. The construction is much simpler than that of the absolute encoder, but more hardware and software is required to interpret the output signals.

The pattern of the disc is seen in Figure 6.7(a). Sometimes the outer ring contains a single mark for generating a synchronizing pulse when it passes the photocell P_1, indicating the 0 or reference position. Movement in either direction from this position generates a stream of pulses from the photocells P_2 and P_3 as the pattern of the two inner rings passes them. Position can be determined by counting the number of pulses, and incrementing or decrementing a counter according to the direction of travel. The direction is determined from the relationship between the two signals P_1

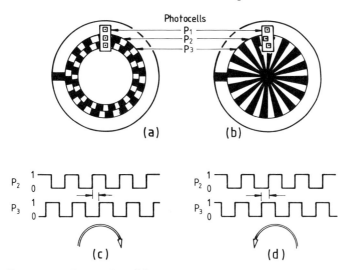

Figure 6.7 Incremental encoder: (a) disc pattern (b) alternative photocell arrangement (c) clockwise signals (d) anticlockwise signals.

and P_2, whose corresponding patterned rings are displaced by one quarter of a cycle. The same effect is achieved by the construction shown in Figure 6.7(b), where the photocells, rather than the patterns, are displaced by the same distance.

Figure 6.7(c) shows the outputs from each photocell for clockwise rotation. The alternating signal from P_2 leads the P_3 signal by one quarter cycle, indicating a clockwise rotation. Figure 6.7(d) shows P_3 to lead P_2 by one quarter cycle, indicating rotation in the other direction.

Position-measuring systems

The resolution of position-measuring devices has been mentioned, and it is appropriate here to define the resolution of a *device,* and to distinguish it from the resolution of the position-measuring *system.* The resolution of a position-measuring device is the smallest movement it can detect. Incremental encoders are available which generate 2 000 000 pulses per revolution, giving a resolution of 0.648 seconds of arc. A high-resolution absolute encoder typically has 1024 segments on the disc, giving a resolution of 21.1 minutes of arc. Resolvers, being analogue in nature, have infinitely fine resolution in theory, but in practice the resolution is limited by the circuitry processing the phase-shift signal. In general, increasing resolution is accompanied by increasing cost.

Demanded positions, movements or velocities in a robot system are specified in automatic running by binary numbers retrieved from a digital memory or calculated by the processor. It is therefore common to refer to the *n-bit resolution* of a robot position-measuring system. Thus, if the output of the converter in the resolver system seen in Figure 6.5 is 14-bit, then the 14-bit resolution is equivalent to $360/2^{14}$ degrees, i.e. 1.32 minutes of arc.

The resolution of a control system is restricted by the resolution of the measuring system upon which it relies for feedback. The choice of measuring system and specification of control system requires the designer to balance performance against price for each element of the control system. Such compromises result, for example, in many robot position-measuring systems using rotary devices to measure linear movement, or using gearing to improve resolution over the direct-drive situation. Figure 6.8(a) shows linear movement of a link on a manipulator obtained by a motor driving a leadscrew arrangement. The encoder measures the movement of the motor, and has a 12-bit resolution. If the pitch of the leadscrew is 4 mm, the linear resolution of the system is 0.977 μm. Regardless of whether an absolute or an incremental encoder is used, it is seen that when the system is initially powered, the computer cannot determine position since the encoder revolves more than once through the full travel of the joint. A known starting position must be obtained by driving the robot manually, or by driving each joint automatically to one extreme of movement, indicated by operation of a limit switch in the form of a proximity detector (Figure 6.8(b)).

The speed of a robot is often limited by the measuring system used.

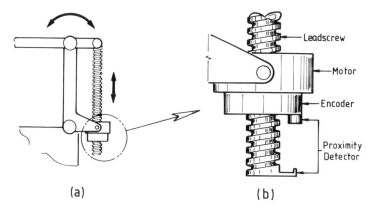

Figure 6.8 Position measurement on ball screw drive: (a) general arrangement (b) encoder and proximity-detector mounting.

Consider the encoder in Figure 6.8(a) to be a 12-bit incremental device, with a maximum tracking rate of 500 000/s. The motor must not be driven faster than 122 rev/s, which is equivalent to a linear speed of 488 mm/s, otherwise pulses may be missed and a false position measurement computed.

If speed must be sacrificed for resolution, the best of both worlds may be achieved by combining coarse and fine resolution systems (Figure 6.9(a)) on one axis. Here one encoder measures the joint position directly while another similar device, geared up 16:1, adds four bits to the resolution without limiting the tracking speed. Hybrid systems can be cost-effective, and one example is shown in Figure 6.9(b), where a resolution of 7.91 seconds of arc is obtained using a 14-bit digital resolver system (which

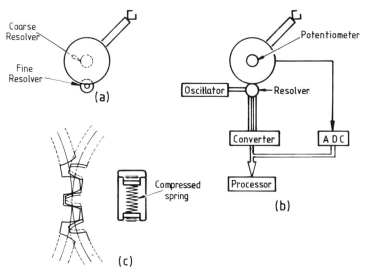

Figure 6.9 Position-measuring systems: (a) direct drive to coarse resolver and gearing to fine resolver (b) anti-backlash gear (c) hybrid potentiometer and resolver system with 1:10 gearing.

on direct drive could only achieve a resolution of 1.32 minutes of arc). The resolution of the potentiometer system is sufficient to inform the processor which revolution the resolver is on, actual position measurement being calculated from the digital output of the converter.

It was stated earlier that a position control system cannot have a better resolution than its measuring system. Resolution is also impaired by the deficiencies of the mechanical and electrical processes in a system. For example, when gears are used, backlash is a potential source of deterioration, and the anti-backlash gear (Figure 6.9(c)) is an important component in these circumstances. In the diagram, the gaps causing backlash are exaggerated to illustrate the effect. Regardless of the direction of rotation, the teeth of the driver are always kept against the same side of the teeth of the driven gear by a second driven gear, which is not fixed to the shaft but held to the first driven gear by a spring. The spring torque depends on how much the anti-backlash gear is 'wound up' during assembly. Low torques are sufficient since only friction in the bearings of the position-measuring device must be overcome; resolvers and encoders do not require any power to drive them at constant speed, save that dissipated as friction in the bearings.

6.3 Velocity measurement

Measurement of velocity is required for closed-loop control of speed and for velocity feedback in position control systems.

Figure 6.10 shows four methods of measuring velocity. The system in Figure 6.10(a) was encountered earlier under the heading of position measurement: this system housed all processing circuitry in an LSI package, and as well as a digital representation of position, a DC output signal

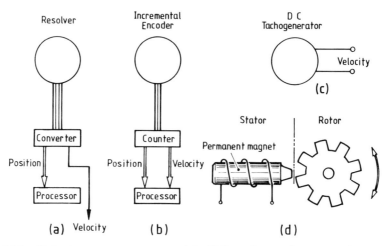

Figure 6.10 Velocity measurement: (a) DC voltage from resolver to digital converter representing velocity (b) digital representation from encoder (c) DC voltage from tachogenerator (d) variable-reluctance tachogenerator.

proportional to velocity is provided. In continuous closed-loop control systems the signal can be used, with suitable amplification, for velocity feedback. In digital control systems, the continuous signal may be processed by an ADC, or it may be calculated by the processor from the rate of change of position.

The incremental encoder (Figure 6.10(b)) in its usual form sends information to a counter which determines the direction of travel from the phase relationships of the signals, and counts position up or down accordingly. The frequency of the pulses is proportional to speed. For example, a pulse frequency of 2000 Hz from a 1000 pulse/rev encoder indicates a speed of 2 rev/s. The counter is therefore able to output digital representations of direction and speed which together constitute a velocity signal. Again, as with any microprocessor system, velocity can be read directly from a suitable device (Figure 6.10(b) being an example) or it can be computed from position changes.

The term 'tachogenerator' applies to any device which generates an electrical output related in a defined way to the speed of a mechanical input. The DC tachogenerator (Figure 6.10(c)) is simply a DC generator coupled directly, or by gearing, to the joint whose velocity is to be measured. The generated voltage is approximately proportional to the speed, and the direction of rotation is indicated by the polarity of the output. The linearity of a tachogenerator is a measure of how closely the output follows the relationship of proportionality between speed and output voltage; high-quality tachogenerators with a linearity of 0.1% are available.

Long lines from the tachogenerator to the signal-processing circuitry are unavoidable on a robot, and in some cases the resulting voltage drops can cause problems. The AC tachogenerator, like the resolver, has the advantage of generating an output which does not deteriorate with voltage drop or a small amount of noise. However, only one AC signal is outputted by an AC tachogenerator, and although the amplitude of the signal varies with speed, it is the frequency of the signal which is measured, since it is related linearly to the speed of the input. This type of tachogenerator, properly called the variable reluctance tachogenerator (Figure 6.10(b)) generates the output from windings round a permanent magnet mounted in the stator. The rotor consists of a toothed wheel of ferromagnetic material, which passes in close proximity to an iron pole extension of the magnet. As the wheel revolves, the reluctance of the magnetic circuit of magnet, air gap and wheel varies cyclically as each tooth passes the magnet. A back e.m.f. is thus induced in the windings, an AC signal of frequency $NR/60$ Hz, where N is the number of teeth on the wheel and R is the speed of rotation in rev/min.

6.4 Force measurement

This chapter is concerned with sensing internal to the machine: measurement of quantities necessary to provide feedback in servosystems. Most robot servosystems control position and velocity, and the devices used for providing feedback of these quantities have been considered. Control of the force being applied by the end-effector is a useful addition to a robot's abilities, and this is often achieved by measuring the force and using the information to alter the position of the robot's end-effector, thereby bringing the applied force to a desired value. The force could be measured by a device external to the robot. It could also be measured by a device on the wrist, or on the arms of the manipulator, or by measuring torques and forces at the machine joints. Building-in force measurement of this kind makes it possible to provide the robot with force control, allowing the user to request applied forces by high-level commands in whatever language is used for teaching or programming the robot. This section considers the hardware used for force measurement, and the interpretation of signals to provide information about applied forces.

Strain gauges

The electrical resistance of a strain gauge varies with its length, which in turn varies with the applied force. By fixing the gauge to the edge of a beam (Figure 6.11(a)), which could be the manipulator arm or a link in a force sensor mounted on the wrist of the robot, the force or torque acting on the beam may be inferred from the change in resistance of the gauge. The performance of a strain gauge is quantified by the gauge factor G, which is defined as

$$G = \frac{\Delta R/R}{e}$$

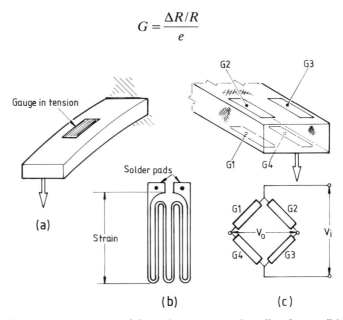

Figure 6.11   Force measurement: (a) strain gauge on bending beam (b) foil pattern of strain gauges (c) strain gauge bridge for selective measurement.

where ΔR is the change in resistance, R is the unstrained resistance, and e is strain.

The popular copper/nickel/manganese alloy used for strain gauges gives a gauge factor, in common with most metals, of about 2.1. This alloy is suited to the application since its electrical and mechanical properties are relatively unaffected by temperature changes. A useful gauge offers an unstrained resistance of 120Ω, which would necessitate a long length of wire of narrow cross-section: this can be achieved by etching out the length in a zig-zag pattern from foil, which is cemented to the surface to be monitored, usually with an epoxy adhesive (Figure 6.11(b)).

Semiconductor materials are also used for strain gauges; p-doped silicon gives a gauge factor of typically 100, while a factor of -100 is observed with n-doped silicon. (A negative gauge factor indicates that the resistance of the gauge decreases as strain increases.) The high gauge factor indicates that these materials are very sensitive to strain, but they have the disadvantage that their resistance, and the gauge factor itself, are susceptible to temperature changes.

It is desirable that the measuring system is sensitive to bending of the beam in the direction of interest, but is insensitive to bending or twisting in other directions. A variety of strain gauge configurations have been devised to suit particular applications; the four-gauge bridge is useful (Figure 6.11(c)). The output of the bridge is

$$V_0 = \left[\frac{R_3}{R_2 + R_3} - \frac{R_4}{R_1 + R_4} \right] V_i$$

The factor in brackets changes as the beam bends in the direction of interest, but does not alter with bending in a direction perpendicular to this, or with twisting.

Piezoelectric polymers

Strain gauges have long been the traditional method of force measurement; they have been successful in many robotic applications. Piezoelectric devices, however, are now offering competition to strain gauges.

It is a well-known phenomenon that atoms in a crystal structure are displaced when a force is applied to the material, the displacement being proportional to the force. Furthermore, a piezoelectric material acquires an electrical charge proportional to the displacement, and therefore proportional to the applied force. By bonding metal electrodes to opposite faces of a film of piezoelectric material (Figure 6.12(a)) a piezoelectric force sensor is formed, the electrodes being necessary to collect the charge.

Electrically, a piezoelectric force sensor is a charge generator in parallel with a capacitor (Figure 6.12(b)). The current generated is proportional to the rate of change of the electrical charge, so the piezoelectric force sensor generates a current proportional to the rate of change of the applied force. For alternating force inputs, the AC output current is easily interpreted to

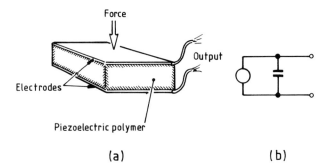

Figure 6.12   Piezoelectric force sensor: (a) construction (b) equivalent circuit.

reveal the amplitude and frequency of the input, but a steady force results in zero output, since there is no gradient to the input. However, we shall see in Section 6.5 that it is possible to integrate the output to give information on steady force inputs.

Many synthetic polymers exhibit the piezoelectric effect, and these materials are especially useful since they can be made in forms suitable for robot sensor applications. For example, their flexibility allows their use in external force sensors such as artificial skin (Chapter 8).

Polyvinylidene fluoride (PVF_2) has proved suitable for robotic applications, since it exhibits a strong piezoelectric effect and is mechanically sturdy. This polymer is available commercially in a variety of forms, and one method of constructing sensor material from it has been reported in detail by Dario *et al.* (1983):

(1) PVF_2 film, which is not piezoelectric in its raw form, is stretched to four times its original length. This causes the crystalline phase of the original film to be altered to a piezoelectric state.
(2) An annealing treatment encourages dimensional stability.
(3) Electrodes are deposited on the polymer in a vacuum. Materials used are aluminium, nickel/chromium and chromium/gold.
(4) A strong electric field is applied to the polymer by maintaining a voltage between the electrodes; this 'thermal probing' process stabilizes the molecular structure of the material.

Force and torque measuring systems

It has been established that a force may be measured by monitoring strain of strain gauges, or more directly by piezoelectric devices. With respect to internal force measurement in robots, there are two categories of system:

• Those designed to measure a force or torque in a particular direction relative to the wrist or gripper.
• A more general measuring system, the most comprehensive giving information about forces and torques about all three orthogonal axes–a force and torque sensor with six degrees of freedom.

Consider the first type of system. Usually two problems must be solved: the system should have good sensitivity in relation to the force or torque of interest, but also the system should be insensitive to forces or torques in other directions.

Figure 6.11(c), for example, shows how strain gauges may be configured for selective measurement of bending, and a simple set-up like this can be used to measure the bending of a link of the manipulator. Knowing the geometry of the robot at the instant of measuring, it is therefore possible to calculate the weight of an object held by the gripper. Initial calibration of the system is necessary, to establish how much deflection is accounted for by the weight of the components of the manipulator itself.

Figure 6.13 shows a gripper linked to the wrist by three force transducers, each of which measures force along the axis of the wrist only. If it can be assumed that no torques are being applied to the gripper, which would be true if the centre of gravity of parts always coincided with the tool centre point (TCP), for example, then the three measured forces constitute enough information to calculate the forces at the TCP. The transducers are disposed at 120° intervals round a circle of radius R, whose plane is a distance D from the TCP. If forces F_a, F_b, F_c measured at the transducers, are considered to be positive when compressive then the forces being applied by the gripper at the TCP are given by

$$\begin{bmatrix} F_x \\ F_y \\ F_z \end{bmatrix} = \begin{bmatrix} 0 & -0.866R/D & 0.866R/D \\ 1 & 1 & 1 \\ R/D & -0.5R/D & -0.5R/D \end{bmatrix} \begin{bmatrix} F_a \\ F_b \\ F_c \end{bmatrix} \qquad (6.2)$$

It is important to stress the limitations of this type of sensor; not only can it not measure torques, but any moment applied to the gripper would reflect as forces in the transducers, and make eqn (6.2) invalid. The value of

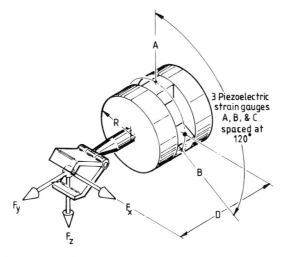

Figure 6.13 Three-axes force sensor using three wrist-mounted strain gauges.

this type of system is that it is simple, and therefore much less expensive than more comprehensive devices. Applications exist where there is only point contact between the tool and workpiece; here torques cannot be transmitted to the wrist and a selective system is sufficient.

Six-axis torque and torque sensors ('load-cells'), are available commercially. A typical unit has the transducer packaged in a sealed unit, making it suitable for industrial environments, and includes a microcomputer unit to resolve the measurements into forces and torques in three orthogonal axes, and output the information via a standard interface, e.g. RS 232. The demand for force measurement in robots has encouraged the development of sensors designed for this area (Gaillet and Reboulet, 1983). These will receive further attention in Chapter 8.

6.5 Analogue-digital-analogue conversion

In Chapter 4 it was shown that many actuating devices are driven by analogue signals, e.g. DC current to an electrohydraulic servovalve. Digitally driven devices are also met, e.g. stepping motors. In Chapter 5, control systems handling both analogue and digital signals were examined, and earlier in this chapter both types of signals were seen to arise in position, velocity and force measurement systems. Hence in nearly all computer-controlled systems instances arise where it is necessary to convert signals from one type to another. The devices which accomplish this operation are called analogue-to-digital converters (ADCs) and digital-to-analogue converters (DACs) (Hnatek, 1976).

In this section the construction and performance of converters is covered after considering some basic components essential to their operation. DACs are detailed before ADCs, since some types of ADC use an internal DAC for feedback purposes.

Operational amplifier

The functions of amplification, summing, comparison, integration and ramp generation are required in DACs or ADCs. All may be achieved by means of an operational amplifier. The operational amplifier (op-amp), a very high-gain differential amplifier, is always used with negative feedback (Figure 6.14(a)). In this mode it acts as an inverting amplifier:

$$V_{out} = -\frac{R_f}{R} V_{in} \qquad (6.3)$$

An extension of this circuit forms a summing amplifier (Fig. 6.14(b)), whose output is

$$V_{out} = -R_f \left\{ \frac{V_1}{R_1} + \frac{V_2}{R_2} + \cdots + \frac{V_n}{R_n} \right\} \qquad (6.4)$$

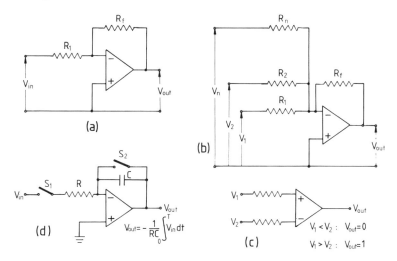

Figure 6.14 Operational amplifier circuits: (a) inverting amplifier (b) summer (c) integrator (d) comparator.

The function of a comparator is to compare two voltages and to indicate which is the greater. The basic circuit for using an op-amp as a comparator is shown in Figure 6.14(c). When the voltage connected to the positive terminal exceeds the voltage on the negative terminal, the output rises to a maximum voltage, considered as ON or a logic 1. If the relationship between the voltages is reversed the output is OFF, logic 0. The response time of an op-amp comparator may not always be satisfactory, in which case more specialized devices with a faster response are used.

Turning now to integration, we note from Figure 6.14(d) that the op-amp requires a capacitor in the feedback line. The device is reset by opening switch S_1 and closing S_2. To perform integration, S_2 is opened and S_1 closed. The output voltage after time T is given by

$$V_{out} = -\frac{1}{RC} \int_0^T V_{in}\, dt \qquad (6.5)$$

A ramp, or steadily increasing voltage, is obtained from the integrator configuration by applying a steady known reference voltage $-V_r$ to the input; eqn (6.5) then becomes

$$V_{out} = \frac{t}{RC} V_r \qquad (6.6)$$

Digital-to-analogue converters

The purpose of a DAC is to convert a digital signal to an analogue voltage whose value is proportional to the binary number represented by the input. Transmission of a digital number of, say, eight bits, requires eight lines, each of which may be ON or OFF, i.e. 5 V or 0 V. The summing amplifier may be

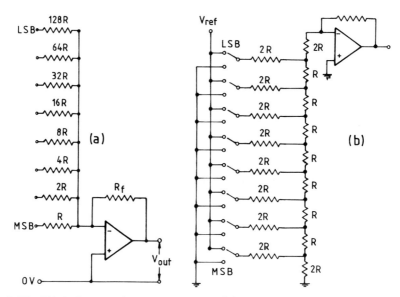

Figure 6.15   Digital to analogue converter: (a) summing network (b) R-2R ladder
network.

used directly as a DAC. Equation (6.4) specified the operation of the
summing amplifier, and here all input voltages are equal to V, therefore

$$V_{\text{out}} = -R_f\left(\frac{1}{R_1} + \frac{1}{R_2} + \cdots + \frac{1}{R_8}\right)V \qquad (6.7)$$

Each successive bit in the digital word has half the analogue value of the
previous bit, so by weighting each bit with an appropriate resistor the
summing amplifier is a valid DAC (Figure 6.15(a)).

   If the signal from the most significant bit (MSB), bit 7, is input through
a resistance R, then bit 6 should have a resistance $2R$, bit 5 a resistance $4R$,
and so on, doubling the resistance each time until at bit 0 the resistance is
$128R$. The output is now truly proportional to the digital value of the input:

$$V_{\text{out}} = -R_f\left(\frac{\text{MSB}}{R} + \frac{\text{BIT6}}{2R} + \cdots + \frac{\text{LSB}}{128R}\right)V \qquad (6.8)$$

A limitation of this circuit is that the correct weighting of each bit relies on
the resistors having values in exactly the required ratios, a requirement not
easily met using discrete components or integrated circuits.

   Many existing DACs use this technique, but to overcome the difficulty
of obtaining a wide range of resistors, when the accuracy of their resistances
is critical to the accuracy of the converter, the $R-2R$ ladder (Figure
6.15(b)) was developed as a suitable input network for this application. An
eight-bit converter requires 27 identical resistors (making $2R$ from two
resistors in series), and this is easier to achieve accurately with discrete
components or an integrated chip. A reference voltage is split by the ladder

network into a series of voltages, each one half the value of the previous one, and the bit signals are used to switch in or earth an input line.

Analogue-to-digital converters

Two types of ADC, dual-ramp and successive approximation, will be described, being typical examples of devices used to achieve this conversion.

In a dual-ramp ADC (Figure 6.16(a)) an integrator is used to generate a ramp from the unknown voltage for a fixed length of time. The charge across the integrator is then dissipated by applying a negative ramp of known gradient, since it is derived from a reference voltage, and the time taken to reach zero voltage recorded. The control logic procedure, on receipt of SOC (start of conversion) is:

(1) Switch OFF EOC (end of conversion)
(2) Reset counter to zero
(3) Switch integrator to V_a and start ramp generation
(4) Wait for signal from comparator to go OFF
(5) Send one pulse to the counter
(6) If the overflow signal from counter is OFF, loop back to step 5.
(7) Switch integrator to $-V_r$
(8) Send one pulse to the counter
(9) If the signal from comparator is OFF, loop back to (8)
(10) Signal EOC

Figure 6.16(b) shows that for a given input the integrator reaches V_1 at

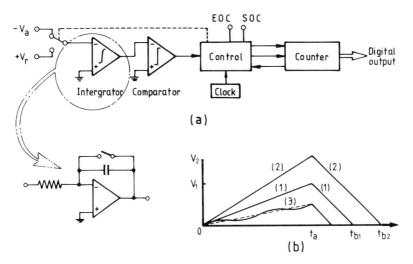

(a)

(b)

Figure 6.16 Dual-ramp analogue to digital converter: (a) structure (b) typical ramps.

time t_a. Since the interval from time 0 to t_a is the time taken for the counter to overflow from 0,

$$t_a = 2^n t_c \qquad (6.9)$$

for an n-bit converter, where t_c is the interval between clock pulses.

Equation (6.5) gave the voltage attained by an integrator; in this instance

$$V_1 = \frac{V_a}{RC} 2^n t_c \qquad (6.10)$$

The same equation applies to the negative gradient ramp, but in this case the slope is different since a different input voltage is applied:

$$V_1 = \frac{V_r}{RC} N t_c \qquad (6.11)$$

where N is the number of pulses registered by the counter. Equations (6.10) and (6.11) yield

$$V_a = (V_r/2^n) N \qquad (6.12)$$

Although the counts registered during the positive ramp time t_a and negative ramp $(t_{b1} - t_a)$ depend on the value of t_c, this quantity does not appear in eqn (6.12): the constant of proportionality relating N to V_a does not vary with t_c. It is also significant that RC does not appear in the equation. For accuracy, t_c and RC should be constant during the conversion period, but their actual values do not affect the converter accuracy. This is why the dual-ramp converter is widely used when accuracy is required; 20-bit resolution is available.

The conversion time of the dual-ramp ADC depends on the value of the input. Reference to Figure 6.16(b) shows that since the second ramp always has the same slope, the time taken to dissipate the voltage, $(t_{b1} - t_a)$ for curve (1) or $(t_{b2} - t_a)$ for curve (2), depends on the voltage attained by the integrator, which in turn depends on V_a itself. Conversion times are similar to those for the single-ramp type.

The operation of the dual-ramp ADC, because of the integration involved, does not digitize an instantaneous analogue value, but indicates the average value over the time period of the first ramp; t_a in Figure 6.16(b). The converter is thus not oversensitive to noise, and if the frequency of possible noise in a signal is known it is good practice to make t_a a multiple of the cycle time, thus effectively rejecting the alternating component of V_a. (See curve (3) in Figure 6.16(b)).

When higher conversion speeds are required, the successive approximation ADC may be used. Figure 6.17(a) shows the structure to be similar to the counter ramp type, but the control logic is different. At each successive step the converter determines if V_a is in the top or bottom half of a selected band of quantization levels by comparing it to an appropriate V_c. The

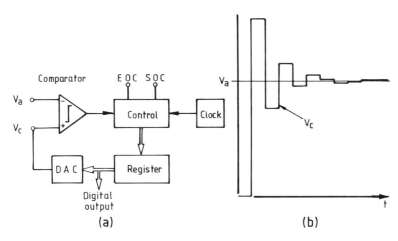

Figure 6.17 Successive approximation analogue to digital converter: (a) structure (b) with input voltage of 1.645, converter sets register to 1.66 V.

selected half is again split into two equal parts so that an n-bit successive approximation ADC reaches EOC in n clock periods. The procedure on SOC is:

(1) Set the MSB in the register to 0, all other bits to 1
(2) If the comparator signal is OFF, change the MSB to 1
(3) Repeat steps (1) and (2) for each bit in descending order
(4) When the LSB has been set, signal EOC

An example will illustrate how this process forces the contents of the register to the value V_a. An eight-bit converter, with maximum input 5.1 V and quantization interval 0.02 V, receives an input $V_a = 1.645$ V. The first step sets the register to

$$0111\ 1111_2 = 127_{10} \quad \text{giving} \quad V_c = 2.54\text{ V}$$

The comparator output is ON since $V_c > V_a$, so the MSB remains at 0. Setting bit 6 to 0 gives the register contents as

$$0011\ 1111_2 = 63 \quad \text{giving} \quad V_c = 1.26\text{ V}$$

The comparator goes OFF since $V_c < V_a$, so bit 6 is changed to 1. The complete process, illustrated in Figure 6.17(b) is as shown in Table 6.2.
This example also illustrates the fact that a successive approximation ADC does not necessarily reach the value nearest to the input. For instance, here the nearest possible reading to 1.645 V would be 1.64 V, but the converter's decision is 83 in the register, which corresponds to $V_c = 1.66$ V. The digitized value is always, however, within 1 LSB (one quantization interval) of V_a.

The conversion time in this device is related to its resolution and does not depend on the value of V_a. Conversion times of 1 μs and resolutions of 14 bits are possible.

Table 6.2

		Register V_c			
Step	*Binary*	*Decimal*	*V*	*Comparator*	*Decision*
1	0111 1111	127	2.54	ON	MSB = 0
2	0011 1111	63	1.26	OFF	BIT 6 = 1
3	0101 1111	95	1.90	ON	BIT 5 = 0
4	0100 1111	79	1.58	OFF	BIT 4 = 1
5	0101 0111	87	1.74	ON	BIT 3 = 0
6	0101 0011	83	1.66	ON	BIT 2 = 0
7	0101 0001	81	1.62	OFF	BIT 1 = 1
8	0101 0010	82	1.64	OFF	BIT 0 = 1
	0101 0011	83			

Comparison of ADCs

The choice of an ADC involves balancing cost against conversion speed and resolution required for a particular application. The possibilities may be compared by listing them in order of increasing cost:

Counter ramp: typical conversion time 1 ms (variable): needs steady input.

Tracking: continuously available output: conversion time needed for large changes in input.

Single ramp: typical conversion time 1 ms (variable): up to 12-bit resolution: subject to temperature drifts.

Dual ramp: typical conversion time 1 ms (variable): up to 20-bit resolution; averages input.

Successive approximation: typical conversion time 1 μs (constant): up to 14-bit resolution.

6.6   Other interfacing problems

Communication between elements of a robot control system requires conversion processes to make the signals from the transmitting device compatible with the receiver. We have seen how conversion between analogue and digital signals is achieved. This section considers two other aspects of interfacing: conversion from voltage to current, and isolation.

Voltage-to-current conversion

Although voltage signals are commonly met in control systems, they are not suitable if transmission over long distances is required. Under these circumstances resistance of the wire itself may be significant enough to introduce a voltage drop in the circuit. Current signals do not suffer from this disadvantage. They are therefore preferred, for example, when an

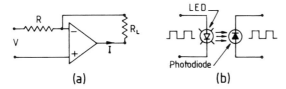

Figure 6.18   Interfacing: (a) voltage to current conversion (b) opto-isolation.

electrohydraulic servovalve mounted on a manipulator is being controlled via a long line from a control cabinet. Figure 6.18(a) shows how an operational amplifier may be used to convert voltage to current. On the assumption that the amplifier gain is high, and the load impedance R_L is low, the circuit obeys $I = V/R$ to a high degree of approximation.

Opto-isolation

Opto-isolators are used to protect a circuit from potential incoming voltage surges. A typical isolator (Figure 6.18(b)) consists of a light-emitting diode (LED) and a photo diode. The LED emits pulses of light in response to input voltage pulses, and these light pulses in turn stimulate the photo diode to emit voltage pulses to the receiving circuit. Thus any robot control circuit receiving signals from an external device may be protected from the accidental connection of damaging voltages by placing an opto-isolator in the input line.

6.7   Recapitulation

Closed-loop control necessitates measurement of the controlled parameter to provide a feedback signal, and a variety of analogue and digital techniques to measure position, velocity and force have been described. Devices such as the resolver, whose output is not adversely affected by the distance between transducer and associated circuitry, are of special importance to robots. ADCs and DACs are often required for computer-controlled systems, and examples have been described. Voltage-to-current conversion and opto-isolation have also been considered.

Chapter 7

Computer control

7.1 Introduction

The development of the computer has been a key element in the evolution of the modern industrial robot. Some of the functions performed by a robot computer control could be undertaken by other specialized devices, for example the pneumatic cascade circuit illustrated in Chapter 5, to implement an event-based sequence, but the overall flexibility required of a robot involves functions which only a computer can perform. Some of these are listed below:

- Control of position, velocity and possibly acceleration and force of the end-effector is required.
- Control of the path taken by the end-effector is often necessary.
- A sequence of operations has to be carried out automatically; often involving interaction with other equipment.
- Decisions have to be taken so that operations may be carried out conditional upon sensory information.
- Facilities are required for an operator to teach the robot and subsequently to amend or add to the instructions.
- There has to be some means of memorizing the task.

A microcomputer, which is a single microprocessor system, cannot cope with all of these tasks. For robot control the computer must contain sufficient processors to provide those functions not undertaken by dedicated devices. This chapter considers how the computer deals with the automatic control of sequence, position and velocity, methods of teaching, software—

the sets of instructions which execute the tasks—and hardware—the construction of the components which make up the computer.

Path control

Different strategies are used to control the path taken by the end-effector during movement from one point in a program to the next. These strategies are of importance during automatic operation, and also in the teaching situation.

Point-to-point (PTP) control requires only the coordinates of start and destination points to be stored. Controlling the speed of the actuators (Section 7.2) helps to smooth the path taken by the end-effector during PTP control, but the actual path is not easily predicted.

Continuous-path (CP) control is required when the path traversed by the end-effector is of importance to the task. The path may be defined in an off-line generated program, or it may be taught directly to the robot.

By using interpolation a PTP control can also move an end-effector along a controlled path: the required incremental motions are calculated from the coordinates of the start and destination and other information such as velocity, entered at the teaching or programming stage.

The CP control of, for example, a painting robot, reproduces the path traced by the spraygun at the teaching stage, when the gun is manually led through the painting operation.

Operating modes

A robot control computer can operate in any of the following modes: off, pre-manual, manual, teach, and automatic. It should be noted that the names given to these modes may vary from machine to machine.

In an *off* mode the computer is inactive, but data stored in memory is preserved if a back-up supply, such as a battery, is provided. In the *pre-manual* mode the computer is operational, but it is not necessary to have power supplied to the manipulator since only data transfer takes place.

The *manual* mode is selected when direct manual control of individual axes is required for setting up and maintenance.

As the name implies, the *teach* mode is used to teach the computer sequences of movements and other operations which will be performed ultimately in the automatic mode. *Teaching* is the umbrella title for methods of entering instructions into the control computer via the robot. By implication, when teaching, the robot is not available for automatic operation. Programs for a robot may be compiled using a separate computer, and this is referred to as off-line programming. On-line programming is, however, a part of the teaching function, since the computer must be taken out of automatic operation and put into a teach mode to accept information, usually from a keyboard.

7.2   Automatic operation

Task execution in the *automatic* mode involves a hierarchy of functions. The simplest, for example, could be a single time-based sequence of moves, performed continuously, with the coordinates of the points in the sequence being transmitted to position-control systems dedicated to each joint. More realistically, an industrial task requires different actions, conditional on sensory information, and when many sensors are involved a structured system is needed to monitor information and to implement the appropriate action.

The hierarchical structure of such a system is exemplified by the robot control system developed by the U.S. National Bureau of Standards (Albus *et al.*, 1983; Haynes *et al.*, 1984). Their control system is divided into a hierarchy of levels (Figure 7.1), each level consisting of three interconnecting modules: sensory processing, world model and task decomposition. The system accepts high-level commands in the form of simple instructions which define complex tasks, e.g. assemble 10 type B engines. These are then broken down into a sequence of lower-level commands, each of which is further decomposed until the lowest level of the hierarchy is reached. This level generates elementary commands for direct transmission to the robot, such as required set-points for robot machine controllers. Actions and decisions in a complex task are often dependent on sensory data, so this information needs to be processed up through the hierarchy.

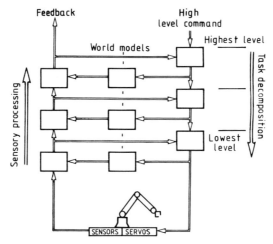

Figure 7.1   Hierarchical robot control system.

Sequence control

The automatic operation of a robot requires control of sequences of events, and the implementation of this type of control for pick-and-place devices using PLCs was covered in Chapter 5. The control requirements of a robot

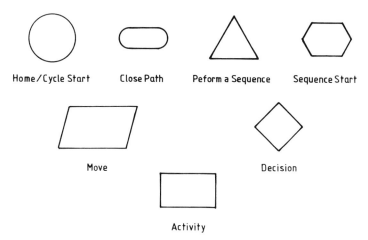

Figure 7.2 Robot program flowchart symbols.

are more comprehensive, involving additionally, for example, programmable position and speed control, and a computer, usually with more than one processor, is required to fulfil all the functions. The elements of the sequences which make up a robot program are best illustrated using the flowchart symbols familiar to the computer programmer (Figure 7.2). Although the language used to symbolize robot functions varies from model to model, the concept of the flowchart is universal.

Consider the robot and turntable arrangement seen in Figure 1.9: the robot is required to perform a welding operation on a part loaded in the jig, while the operator unloads the previously worked part and loads a fresh workpiece. Two aspects of the operation require decisions to be made by the robot: when the table is signalled to turn, the robot must wait until the 180° rotation is completed (indicated by a switch connected to a robot input); and the robot must reject an incorrectly located part (also indicated by a strategically placed switch).

Figure 7.3 shows a flowchart of the necessary program. The main line sequence (0) starts top left, and point 0.002 is the start of the operation cycle. At 0.003 output 3 is turned on (+3), activating a pneumatic solenoid to turn the table. Input 7 is activated by a microswitch when the table has fully turned, so the robot constantly interrogates the input until this condition is satisfied. Input 9 is connected to a switch which is made only if the part is located correctly in the jig on the side of the table presented to the robot; if location is correct sequence 1 is performed. In sequence 1 the robot approaches the part, turns on the torch (+T) and executes the required tasks, taking the sequence to point 1.010 where the torch is turned off (−T), the robot withdraws clear of the table and returns to point *B,* i.e. point 0.006. Had the part been loaded incorrectly the main line sequence would have moved directly to 0.006 without welding. The remainder of the sequence is a repetition of the welding task on the other side of the table, after which the program returns to cycle start, i.e. 0.002.

This simple example highlights some of the features of a robot program.

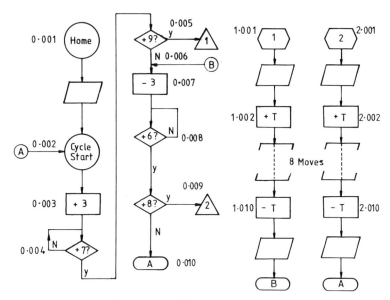

Figure 7.3 Flowchart of robot welding program including turntable operation.

A main-line sequence is a loop of instructions which dictates the order of operations and has conditional statements to take corrective action where necessary. In a well-structured program the physical tasks are contained in subsequences making the order of operations easier to follow, and therefore easier to amend.

PTP control

Earlier chapters have shown how world coordinates can be transformed to machine coordinates for a given manipulator, and how a continuous or digital control system for each joint can achieve the desired positions in world coordinates. This section is concerned with methods of determining the set points for these controllers.

When the actual path between desired positions is not of great importance, PTP control is used; where it is of importance CP control is necessary, or PTP with interpolation.

There are four types of PTP control: sequential; uncoordinated; terminally coordinated, and PTP with interpolation. The method of teaching is common to all types. Only the start and destination are stored in memory, and the path traversed by the tool centre point (TCP) during teaching is ignored. The actual path taken during playback depends on which type of control is in operation.

In sequential control each axis is driven in turn. This form of control is commonly found in robots driven by stepper motors, and controlled by a single microcomputer. The microprocessor unit (MPU) itself is used to generate the pulses which drive the stepper motor, so that it is only possible to drive one axis at a time during teaching and playback. Hence, in moving from one point to another the computer drives the first axis in the sequence to its destination, repeats the process for the second axis and continues until all the axes have attained their new positions. The simplicity of this form of control is offset by the relatively long time required for completion of an overall moment.

Uncoordinated PTP control speeds up the movement by driving all axes simultaneously. A position control is required for each axis, but no control of the speed of an actuator is provided, so that actuators may reach their destinations at different times, and the path traced by the TCP is not therefore easily predictable. Overall speed is often determined by the slowest actuator.

Terminally coordinated PTP control is widely used. The axes are driven at controlled rates, speed control being available, so that all axes reach their destinations simultaneously.

PTP control is used when the trajectory between desired positions of the TCP is not of great importance. It is nevertheless instructive to note how the trajectories differ for different forms of PTP control. Consider a four-degrees-of-freedom jointed-arm robot moving the TCP from the start machine coordinates of

$$(\phi_S, \theta_{1S}, \theta_{2S}, \theta_{3S}) = (45°, 60°, 240°, 330°), \quad \text{i.e. pitch} = 270°$$

to the destination

$$(\phi_D, \theta_{1D}, \theta_{2D}, \theta_{3D}) = (45°, 88.173°, 286.335°, 345.492°), \quad \text{i.e. pitch} = 0°$$

Both points have been chosen to lie on the same vertical radial plane, thereby allowing the whole movement to be easily represented on the two-dimensional page. Figure 7.4 shows the start (S) and destination (D) configurations on the radial plane. The pitch angle of the gripper ξ is defined in Figure 2.16.

The dimensions of the manipulator are $L_1 = L_2 = 1$ m and $L_3 = 0.25$ m,

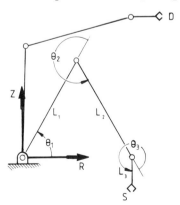

Figure 7.4 PTP control of a jointed-arm manipulator: definition of machine coordinates.

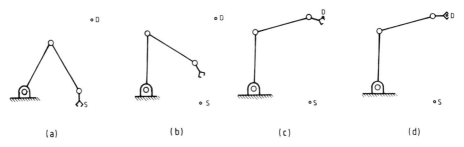

Figure 7.5 Sequential PTP control: (a) start (b) shoulder motion complete (c) elbow motion complete (d) pitch motion complete at destination.

and the required rotations are thus:

$$
\begin{array}{ll}
\text{base} & 0° \\
\text{shoulder} & +28.173° \\
\text{elbow} & +46.335° \\
\text{wrist} & +15.492°
\end{array}
$$

Sequential control is illustrated in Figure 7.5, where Figure 7.5(a) shows the start, Figure 7.5(b) shows the position on completion of the shoulder rotation and Figure 7.5(c) the position after the elbow rotation. Finally the wrist is rotated, thus attaining the destination shown in Figure 7.4(d). If each actuator has a maximum speed of 1.57 rad/s, the complete movement would take 1 s.

Uncoordinated PTP control of the same actuators is illustrated in Figure 7.6. The start is shown in Figure 7.6(a). Assuming that actuator speeds are equal, Figure 7.6(b) shows the configuration when each actuator has turned 15.492°, at which point the wrist movement is complete. The shoulder has to turn a further 12.681°, the resultant configurations being shown in Figure 7.6(c). The final destination is reached (Figure 7.6(d)) when the elbow has turned through 46.335°. If all actuators move with the same speed, the time taken to complete the movement is dictated by the axis requiring the greatest rotation; in this case the elbow. With actuators moving at 1.57 rad/s, uncoordinated PTP control would complete this movement in 0.51 s.

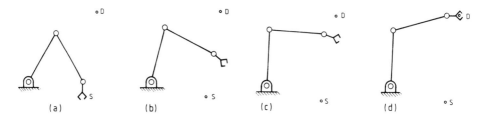

Figure 7.6 Uncoordinated PTP control: (a) start (b) pitch motion complete (c) shoulder motion complete (d) elbow motion complete at destination.

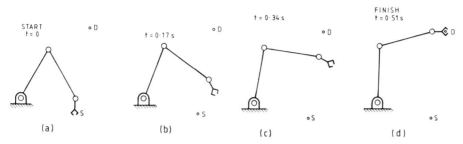

Figure 7.7 Coordinated PTP control: (a) start (b) position after 0.17 s (c) position after 0.34 s (d) destination.

Terminally coordinated PTP control would achieve the movement in the same time as the uncoordinated form, but all actuators are driven at a speed calculated to complete their rotations at the same time. In our example this would require the elbow to be driven at maximum speed, since it has the greatest rotation, and the speed of the other actuators would be calculated so that each completes its rotation in the same time as the elbow. The required rates of rotation are:

$$\begin{array}{ll} \text{base} & 0 \text{ rad/s} \\ \text{shoulder} & 0.96 \text{ rad/s} \\ \text{elbow} & 1.57 \text{ rad/s} \\ \text{wrist} & 0.53 \text{ rad/s} \end{array}$$

Positions after 0.17 s, after 0.34 s and at the destination (0.51 s) are shown in Figure 7.7.

Tool centre point trajectories for differing forms of control are illustrated in Figure 7.8. The results of sequential control are shown in Figure 7.8(a), of uncoordinated control in Figure 7.8(b), and of terminally coordinated control in Figure 7.8(c). It can be seen that, of the three control forms, terminally coordinated control is the only one without discontinuities in TCP trajectory.

Having compared sequential, uncoordinated and terminally coordinated PTP control, we now consider PTP control with interpolation. It should be

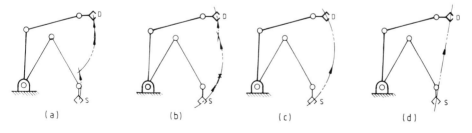

Figure 7.8 Comparison of paths traced by TCP in PTP control: (a) sequential (b) uncoordinated (c) coordinated (d) with linear interpolation.

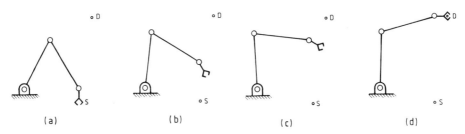

Figure 7.9 PTP control with linear interpolation, programmed velocity of 2 m/s: (a) start (b) position after 0.253 s (c) position after 0.507 s (d) destination reached after 0.76 s.

clear that none of the forms of PTP control already mentioned can give control of the TCP trajectory. If this is required, the computer must calculate the coordinates of intermediate positions on the desired path, i.e. it must interpolate. If the interpolated positions are close enough together, the path taken by the end-effector will come close to the desired path. Straight-line motion is achieved with linear interpolation, and circular interpolation is used when an application requires the end effector to describe a circular motion.

The effect of PTP control with linear interpolation on the previous example is shown in Figure 7.8(d), and Figure 7.9 shows the configuration of the manipulator at the start, movement one-third complete, two-thirds complete, and destination.

The time interval between steps is determined by the speed of the processor and the number of calculations which must be carried out at each step. A Cincinnati Milacron T3 robot, for example, uses AM2901B processors which have a typical instruction execution time of 125 ns (Money, 1982). The T3 is a six-degrees-of-freedom machine, and the necessary 1677 multiplications and 1330 additions are carried out in a time interval of 5 ms (Megahed and Renaud, 1982).

PTP control requires the computer to store coordinate information for every point in a program, and instructions to be carried out at those points. This information is stored in memory, and the memory size of a robot computer control is usually expressed in terms of the number of points which can be held. A typical memory size is 1024 points.

Acceleration and deceleration ramps

A practical manipulator exhibits a degree of flexibility due to bending of the arms and lack of stiffness at the joints, whether hydraulically or electrically driven. Hence attempts to start or to stop large loads abruptly can result in vibrations of the load and manipulator. These can be reduced by accelerating to the required velocity at a controlled rate, (the *acceleration ramp*, typically 4 m/s^2), and decelerating in a similar fashion when approaching the destination.

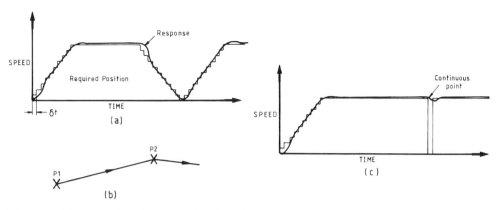

Figure 7.10 Acceleration and deceleration ramps: (a) set-points and response (b) P2 may have no ramps if direction change is small (c) response through a continuous point.

Figure 7.10(a) shows how this strategy would be implemented on a machine using PTP control with linear interpolation. The requirement is to start from rest at point *P1*, move to point *P2*, and thence to another point (Figure 7.10(b)). Instead of demanding the programmed velocity immediately, successive values of the required velocity are increased at a constant rate from 0 to the programmed value. The response of the system, i.e. the actual speed attained by the end-effector, is also shown in the diagram.

If the movement shown in Figure 7.10(b) were part of an arc-welding task, a stop at the point *P2* would cause excessive weld material to build up. For this reason facilities are provided to remove the acceleration and deceleration ramps at a point if required. The point *P2* would be programmed as a *continuous point,* and the required velocity would stay constant except for one time interval during which the computer calculates the path length and number of time intervals for the movement beyond point *P2*. The actual speed of the welding gun through such a continuous point is seen in Figure 7.10(c).

Continuous-path (CP) control

The path control achieved by interpolating between points in PTP control is often called CP control. This section is concerned with another form of CP control, one where the trajectory is determined at the teaching stage by leading the manipulator through the task. In the teach mode, machine coordinates are recorded at regular time intervals (5–100 ms) as the manipulator is moved through the required movements. The computer's task is to use this information to reproduce the original movements as accurately as possible. In the simpler systems, sets of coordinates are read from memory at the same rate at which they were recorded, and used directly as set-points for position control systems.

Unlike PTP control, the number of points in a CP controlled program is directly related to the length of time the program runs; for example, a 5 min program operating on a sampling time of 10 ms has 30 000 points. A typical paint-spraying robot can record 30 min of operation, and this is achieved by storing the information on tape or disc, with portions of the program being dumped to random access memory (RAM) as required.

Tracking

Many industrial applications require a robot to perform a task on a moving workpiece. Two examples are spot-welding car body units on a continuously moving line, and painting workpieces which pass in front of a robot on an overhead conveyor. Satisfactory operation is unlikely to be achieved if the robot is taught whilst the workpiece is moving, since any discrepancy between the speed of the line when teaching and when the robot is an automatic mode would displace the workpiece from the position expected by the robot.

Tracking is the ability of a robot to apply a program, taught using a stationary workpiece, to a situation in which the workpiece is moving. Tracking is achieved by monitoring the position of the part and continuously adding its displacement from the teach position to the desired position of the end effector. The relevant calculations must be carried out in world coordinates, so it is necessary to employ a computer control which has the ability to transform machine to world coordinates and vice versa.

Consider the spot-welding operation shown in Figure 7.11(a). The memory contains programmed world coordinates calculated from joint-angles generated during teaching on a stationary part (Figure 7.11(b)). In the automatic mode, these stored coordinates are used to generate a series of coordinates which constitute the required path of the welding tip. The displacement of the conveyor line from the taught position is constantly monitored, and this displacement is added to each point on the generated

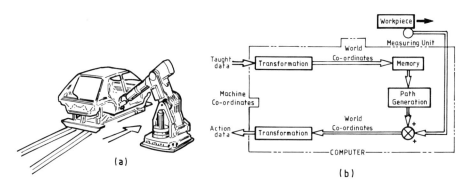

Figure 7.11 Spot-welding on a continuously moving conveyor: (a) general arrangement (b) world coordinates are stored in memory, and amended by position measurement from conveyor before transformation to machine coordinates.

path (Hohn, 1978). Thus the program tracks the line, and if the movement of the line is parallel to the y-axis, as in this case, then only the y-coordinates in the path generation would be modified. The modified path is then transformed in the usual way to generate the required joint-angles for the manipulator.

On-line program alterations

Some current industrial robots have facilities for altering a program during its execution. Generally the structure of the program must remain intact, but parameters such as point coordinates, path velocity or the switching of outputs, may be altered.

Consider a robot vision system which uses a television camera to view objects approaching the robot on a conveyor belt. The camera is connected to a microcomputer which in turn communicates with the robot computer via an RS232 link. The microcomputer vision system has the ability to detect an object and to determine its coordinates on the conveyor. Regardless of the location of the object, the robot's task is to approach it, pick it up and deposit it in a package at a known location. The robot is programmed to seek the coordinates of the object from the external microcomputer, and when these are received by the robot computer they are immediately echoed back to the microcomputer for checking. If they are correct a check signal is transmitted to the robot, and the program is allowed to continue.

7.3 Teaching

This section considers the various systems that enable an operator to teach a robot a task in a step-by-step manner. One of the more common teaching methods requires the end-effector to be driven through the required sequence of movements manually: another involves the use of a model manipulator. In addition, teaching from a remote pendant is possible, and instructions may also be accepted by the computer from a keyboard mounted at the control cubicle. Teaching is not the only method of putting a program into the computer memory; a program can be compiled off-line, i.e. without use of the manipulator or robot computer, and transmitted to the robot afterwards (see Section 7.4). The present section, however, is confined to teaching. Three teaching methods are identified:

- Lead-through: typically a paint-spraying application.
- Walk-through: teaching by driving the end-effector from a hand-held pendant, and programming points, functions, etc., from the pendant and possibly a keyboard.
- On-line programming: all instructions are entered directly into the computer via a keyboard, including the numerical values of co-ordinates.

Teaching methods

The method of teaching a robot depends on its construction and application. Consider an application such as spray-painting, where the requirement is to teach the robot a sequence of movements and the points at which the spraygun has to be switched on and off. This is achieved by lead-through. With the control in the teach mode, the operator grasps the gun mounted in the manipulator and leads it through the painting operation. During the process it is necessary to disconnect the power supplies to the manipulator so that the various axes are easily moved manually. At regular time intervals the resolvers or encoders are scanned by the computer, and their angles and displacements recorded. In the automatic mode these values can be transmitted as set-points to the position control system for each axis.

Counterbalancing must be provided on larger robots in order to enable the manipulator to be handled in this way: alternatively, a teaching arm can be used. The arm is simply a geometrical copy, a full-scale model, of the manipulator itself, but fitted only with the end-effector and position-measuring devices on each axis. This lighter mechanism is then used for the teaching operation, the computer control recording axis positions from the teaching arm and then using the same values to reproduce the movements in the manipulator iself. (Chapter 10 will show how such a mechanism can be used in a tele-operator system.)

In multirobot installations with identical robots only one teaching arm is necessary, the arm being easily plugged in to a selected machine for teaching. This cost-saving principle may be expected to be applied to other expensive pieces of hardware common to more than one robot in a plant and yet only required for teaching, e.g. pendant, keyboard and VDU.

The hand-held pendant is also widely used for teaching robots. It is essential for those applications where an operator is required to drive the end effector accurately to defined positions. In general the pendant has six pairs of controls driving six axes in both directions. The controls, using joysticks or push buttons, may drive each actuator directly. Alternatively, if the computer control has the ability to control the path which the end-effector follows (see Section 5.4), the pendant controls will produce movements in world coordinates or other reference frames when the robot is in the teach mode.

A robot drilling application offers an example of a teaching situation where this facility is necessary (Figure 7.12). The air-powered drill has to be positioned by an operator, using a pendant, to the start of the hole to be drilled, and has to be correctly orientated. When the drill is being positioned, the pendant controls produce movements in world co-ordinates. For example, the OUT button produces a movement in the X-direction, with the other five coordinates remaining unaltered, (decoupled control). When teaching the drill feed, the computer is switched to a hand coordinate system. When this is done, the pendant controls relate to the direction in

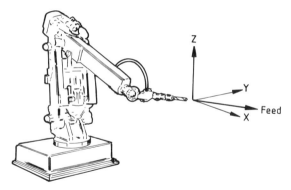

Figure 7.12 Drilling operation: required direction of feed at an angle to all three axes.

which the end-effector is facing, so that pressing the OUT button produces a straight-line movement along the axis of the drill.

During teaching it may be necessary to program a variety of functions. Parameters such as tool velocity may have to be changed during the course of the program, and the operator requires feedback information on the state of the system. The information is displayed in the form of lights or LEDs on the pendant, or on the VDU in the control cubicle. The design of pendants varies considerably, since manufacturers must compromise between the provision of control of all functions and full information display on the pendant, while avoiding an unnecessarily complex pendant. The ideal pendant would be simple to use and yet have all these facilities, possibly providing information visually on a small flat-screen VDU, or audibly using speech synthesizers.

An important safety feature of the pendant is that it operates on the 'dead-man's handle' principle, i.e. motion of the manipulator will cease when all controls on the pendant are released.

Program example

Consider how the short program introduced in Section 7.2 can be taught to a Cincinnati Milacron T3 robot (Cincinnati, 1980a). This machine is walked through using the hand-held pendant shown in Figure 7.13. Program points with or without ON or OFF signals can be transmitted to two tool connections. During the course of teaching, the robot can be driven through moves already programmed with forward and backward buttons. All other functions, e.g. a change of velocity, must be entered at the keyboard on the control cubicle.

During teaching all information about the current point is displayed on the VDU; this includes sequence and point number, function programmed, velocity and world coordinates of TCP. The program may be followed by

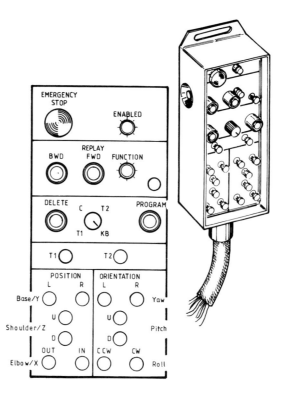

Figure 7.13   Teach pendant of Cincinnati Milacron T3 robot.

listing the point numbers along with the function programmed at each point:

0.001	HOME	(Sequence 0, point 1)
0.002	CYCLE START	(Cycle loops back to here)
0.003	OUTPUT +03	(Switch ON output 3)
0.004	WAIT, UNTIMED, +S07	(Sequence continues after input 7 comes ON)
0.005	PERFORM 1, +S09	(Start sequence 1 if input 9 ON)
0.006	NOP	(No operation)
0.007	OUTPUT −03	(Switch OFF output 3)
0.008	WAIT, UNTIMED, +S06	(Sequence continues after input 6 comes ON)
0.009	PERFORM 2, +S08	(Start sequence 2 if input 8 ON)
0.010	NOP	
CLOSE PATH TO 0.002		(Return to cycle start)
1.001	NOP	(Start of sequence 1)
1.002	+T1A	(ON signal to welding controller)
⋮		Actively welding
1.010	−T1A	(OFF signal to welding controller)
1.011	NOP	

CLOSE PATH TO 0.006   (Return to main line sequence)
2.001   NOP   (Sequence 2 similar to sequence 1
2.002   +T1A

   ⋮

2.010   −T1A
2.011   NOP
CLOSE PATH TO 0.002   but different close path
                                 instruction)

There are two points worth noting. It is good practice to work out the structure of the program before approaching the machine. Having done this, the operator knows, for instance, to program a dummy point at 0.006 (NOP); this point is needed to close the path at the end of sequence 1. Apart from the close path location, sequences 1 and 2 are identical. In fact the welding sequence is only walked through once, when sequence 1 is programmed. Sequence 2 is created at the keyboard by a copying facility, using sequence 1 as the model and the appropriate close path, here 0.002, inserted.

An instruction such as that at 0.005 is a conditional branch; here if input 9 is on, indicating a correctly loaded part, sequence 1 is performed. The instruction is ignored if the input is off.

A computer language is a formal structure of commands which allows an operator to communicate instructions to the machine; by inference, then, any robot has a language, in the sense that the operator must learn the correct method of inputting information to the computer control. Normally computer languages cater for keyboard input only, but in a robot different methods are necessary, e.g. keyboards, teach pendant, teaching arm, to allow instructions to be communicated to the computer.

7.4   Software

The operator of a robot should not be required to have a detailed knowledge of the arithmetical and logical operations carried out in the execution of a task. Hence all practical machines have software on board which enables instructions to be communicated to the robot in a language which the operator can easily understand. The ideal robot would accept verbal instructions in a natural language, and, as we shall see in Chapters 8 and 10, voice recognition systems are now at an advanced stage. Nevertheless, current machines require the operator to use a formal language as an interface between the robot control and the external environment.

The higher the level of the language, the simpler the instructions required of the operator. This, however, is at the expense of the additional effort and computer time required to convert instructions into machine language—the language of the microprocessor. Four broad aspects of language will be considered: objective languages, robot languages, high-level languages, and assembly and machine languages.

Objective languages

Objective robot languages allow the task to be specified in a general form, usually in written natural language form; for example 'assemble 3 pumps type 30'. Objective languages require high computing power to interpret the commands, and advances in artificial intelligence and voice recognition techniques (Levas and Selfridge, 1983) may be expected to lead to robots which may be taught and programmed by spoken natural language (Evard *et al.*, 1982). An example is RAPT (Poppelstone *et al.*, 1978; Ambler *et al.*, 1982) developed at Edinburgh. It works with frames of reference defined relative to objects rather than the manipulator itself.

Commercial languages

The slow speed of general-purpose high-level languages, and their lack of good input and output facilities, has made them unattractive for robot control: special robot languages are preferred.

The teaching methods studied in Section 7.3 involve use of the robot in the actual workplace, along with any other associated plant and prototype workpieces. This may be undesirable in small batch production where equipment could frequently be down for such teaching sessions. Since the set of instructions and data which make up a robot program are stored in digital memory, it is easily transmitted to the robot computer from an external computer. This process is termed *off-line programming* (Tarvin, 1980). It is not to be confused with the facility of loading a program from an external memory device, for in this situation the program would have been written initially using the robot. Off-line programming allows a program to be written in a form compatible with the robot control but not requiring the control or manipulator for the writing process.

Programming, rather than teaching, has the following advantages:

- Use of production equipment is not interrupted.
- Hazards to safety are eliminated since the operator is not inside the manipulator's working volume.
- High-level 'robot languages' can be used to simplify much of the work involved in writing a program.
- Data in computer-aided design (CAD) systems can be accessed directly for use in the program. (See Section 1.9).

Teaching however offers an opportunity to validate the program *in situ*, and errors such as collisions with other equipment, are unlikely to arise. Such errors in programming must be guarded against by complex software, or by the programmer viewing a simulation of the manipulator movements on a VDU.

The purpose of a robot language is to make it easier for the operator to define the task to be undertaken by a robot. Robot languages fall into the following categories, requiring that

- machine coordinates must be specified,
- coordinates of the end-effector must be specified,
- coordinates of the workpiece only need to be defined.

There are many robot languages on the market (Gruver *et al.*, 1983) and the reader is referred to the literature for further details. Table 7.1 includes some of those available today:

Table 7.1

ADA	Volz *et al.* (1983)
AL	Mujtaba (1980), Mujtaba and Goldman (1979)
AML	IBM (1981)
Animate	Kretch (1982)
APT (see MCL and RAPT)	
AUTOPASS	Liebermann and Wesley (1977)
HELP	Donato and Camera (1981)
JARS	Gruver *et al.* (1983)
MCL	Wood and Fugelso (1983)
RAIL	Automatix (1981)
RAPT	Popplestone *et al.* (1978)
RPL	Park (1981)
Sigla	Sigma (1980)
VAL	Carlisle *et al.* (1981), VAL (undated)

High-level languages

High-level languages such as BASIC and FORTRAN simplify the programmer's task of writing and entering programs: they are user-oriented. A high-level language program generates many instructions when translated into machine code. With FORTRAN this is done after the program is written, a process called *compiling,* and the machine instructions, the compiled program, is stored in memory. Languages such as BASIC are stored as a list of high-level instructions; each time the program is run the instructions are *interpreted,* i.e. translated into machine code, line by line; interpreted language programs thus tend to execute more slowly.

When certain operations in a program have to be carried out frequently, they can be speeded up by writing them directly in machine code. Indeed, when operating a robot from a general-purpose microcomputer, the main structure of the program will invariably be written in a high-level language (usually BASIC), whilst frequently-used operations, e.g. control algorithm or coordinate transformations, will be lodged in machine-code subroutines.

Assembly and machine language

Assembly-language programs are written using the mnemonics which represent each MPU instruction. Writing a program in assembly language requires a detailed knowledge of the MPU operation. Allied with each mnemonic is the address of the data to be manipulated. An assembler program is necessary to translate assembly language to machine code before the program may be executed.

It is not the function of this text to teach microprocessor programming. The aim is rather to provide an understanding of the processes involved in translating operator instructions into proper control of the manipulator. In order to achieve this the hierarchy of languages in the system, from high-level languages down, has been followed, leading ultimately to machine language, the codes used by the microprocessor.

The elements of a microprocessor can adopt only one of two electrical states—ON or OFF. They are thus ideally suited to a binary system in which OFF is equivalent to 0 and ON is expressed by 1. A single binary number is termed a *bit,* and microprocessors can process 'words' of 4, 8, 16 or 32 bits. An eight-bit word such as 10101100 is called a *byte.* A byte is transmitted in parallel along eight data lines of the bus connecting the elements of a microprocessor.

Binary numbers suffer from the fact that large strings of bits are needed. The hexadecimal scale gives a more compact presentation by writing numbers to the base 16. The letters A to F represent the decimal numbers 10 to 15, and a hexadecimal number is denoted by the prefix $. A byte requires two digits in hexadecimal form, so the decimal number 138, for example, is equivalent to $8A.

A single MPU has a number of functional elements which communicate with each other to carry out instructions sent to the MPU. An MPU executes a machine-code program by reading and carrying out in sequence the instructions contained in memory. The instruction set gives three- or four-letter mnemonics for each instruction, which aid the programmer when writing a program in assembly language.

7.5   Computer control hardware

The cubicle which houses the computer control contains the many different systems required to fulfil the functions of the machine. This section considers the hardware necessary to perform calculations and logical operations, to control position and velocity, to provide power supplies, and to interact with an operator and other equipment.

Microprocessors

One or more microprocessors make up the central processing unit (CPU) which handles the calculations and logical operations necessary for control of the robot. The 'chip', the more familiar name for the microprocessor, is a thin slice of silicon. A large-scale integrated circuit (LSI) of transistors, diodes and resistors can be fabricated on such a slice. The chip is often enclosed in a plastic or ceramic material, with connections usually taken out to 20 pins along each side, forming the standard 40-pin dual-in-line (DIL) package.

There are two commonly used methods of fabricating microprocessors. Bipolar devices are constructed from transistor–transistor logic (TTL), emitter-coupled logic (ECL) or integrated injection logic (IIL) circuits. These devices are to be found in many robot controls because their very fast operation is necessary for the complex calculations required for robots with controlled path movements. Bipolar circuits are not as compact as others: for example a number of chips may be needed to perform the same function as one metal oxide silicon (MOS) processor.

pMOS and nMOS microprocessor circuits are built up respectively from p and n channel field-effect transistors (FET). Complementary MOS (CMOS) microprocessors use both n-type and p-type FETs, and although slower than their pMOS and nMOS counterparts, they have the advantages of tolerance to power supply voltage variations and low power consumption. For example, the Hitachi HMCS 42/43 (Money, 1982), a pMOS type, consumes 100 mW at 10 V; the HMCS 42C, an equivalent CMOS type, consumes 1.5 mW at 5V.

Microprocessors typically have an operating range of 0°C to 70°C. Since their power consumption is low, heat dissipation is not usually a problem. The heat generated by other items such as visual display screens or power units is of more significance, and a cooling fan may be necessary to prevent excessive temperature in the microprocessor's environment.

Microprocessor associated devices

Associated with the microprocessor are memory devices which store the sequence of instructions to be carried out by the CPU and data operated on or generated by those instructions, and input/output devices through which the CPU receives and transmits signals from and to other systems.

Instructions and data which have to be retained permanently in a particular system are stored in read-only memory (ROM). The microprocessor can only interrogate ROM, i.e. it may read the data or instructions but cannot write into ROM. ROM is non-volatile, i.e. memory is not lost when power is removed from the device.

As indicated earlier, ROMs are fabricated using bipolar or MOS technology, and although bipolar ROMs contain fewer memory locations in a given area than MOS devices, and consume more power, they have an access time typically one-tenth that of MOS devices. ROMs are mask programmed; this process is carried out at the manufacturing stage, when the program is 'burnt' permanently into the chip. Programmable ROMs (PROMs) contain fusible links between circuit elements, and by blowing selected links, a PROM programmer can be used to implement a desired program on the chip.

Erasable PROMs (EPROMs) take the flexibility of PROMs one stage further. The complete memory can be erased by exposing the chip, which is covered by a quartz window, to ultraviolet light for 30 min. The device may then be reprogrammed. An EPROM must be removed from circuit for erasure, but the electrically alterable ROM (EAROM) has the facility of allowing definable portions of memory to be erased by applying the correct voltage to particular pins while the device is in circuit.

Regardless of the method of programming or reprogramming, ROMs contain information which is not destroyed when the machine is switched off. That set of instructions which is specific to the robot control, and does not require changing, is therefore usually stored in ROM. For example, the instructions for calculating joint-angles from cartesian coordinates would be stored in ROM to be available to the CPU at all times. The tool dimension used in such calculation is, however, a variable quantity, depending on the particular tool in use, so this information would be stored in random access memory (RAM).

Information may be read from or written to RAM by the CPU, 'random access' meaning that access to any chosen address may be achieved in a fixed time. RAM is volatile, i.e. its data is only retained while power is on. In general RAM is used to store all the data and programs which have been generated by the operator or the CPU itself.

Static and dynamic RAMs are available. Static RAMs, which are fabricated from TTL circuits, can handle data at a very fast rate. Dynamic RAMs, fabricated using MOS technology, have a smaller power consumption, and may be more densely packed than the static devices, but require additional circuitry to provide a refresh signal. This signal is required at regular intervals because the information is stored in the form of capacitor charges and must be constantly rewritten several times a second since the charges leak away.

In order to function as a robot controller, the CPU must be able to transmit data and to receive data from other systems: two methods are used.

- *Memory mapping* involves the use of temporary memory locations within the main RAM: these 'registers' can in turn transmit or receive data from other devices. The CPU then sees the registers simply as RAM locations and communicates with them accordingly. The term 'poke' refers to writing data to a memory location: to 'peek' a location is to read, but to leave unaltered, the data stored in it.
- *Input/output* (I/O) *ports* allow a more direct communication between the CPU and external devices, referred to as 'peripheral devices'.

Axis control

Chapter 5 described how continuous and digital control systems could be implemented to achieve closed-loop position control of a single robot axis.

The value of the set-point is available in digital form from the CPU, and this value may be used directly by a digital system, or converted to analogue form to provide the input to a continuous system. The complete control system is usually mounted on one printed circuit board (the axis driver board) and a similar system provided for each axis.

Buses

Figure 7.14 shows a microprocessor system consisting of CPU, memory and I/O ports. The CPU communicates with other devices along three sets of wires called buses.

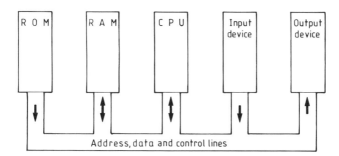

Figure 7.14 Basic microprocessor system: CPU communicates with memory and input/output devices via common bus.

The CPU sends out signals to set memory into the read or write mode, and to synchronize the transmissions of data, along the control bus. The number identifying the memory location to be peeked or poked is carried by the address bus, whilst the data bus handles the actual information.

In a microcomputer the buses are printed on the board carrying most of the components of the machine, but in a robot control it is more important to have all systems easily accessible for testing and maintenance than it is to make the unit small. Therefore the more usual and practical method of constructing the control is to devote separate boards, or 'cards' to each subsystem, e.g. CPU card, axis driver cards. All cards slide into standard racks in the control cubicle, plugging in to the buses at the rear of the rack.

Power Supplies

The source of all electrical, pneumatic or hydraulic power for an industrial robot is the three-phase or single-phase electricity supply. The control cubicle requires a variety of DC voltages for electronics, display lights, inputs and outputs, and also an AC supply for cathode ray tube (CRT) displays and AC inputs and outputs. The required AC voltage (usually 110 V) is obtained by transforming the mains supply: DC voltages are obtained by rectification. Solid-state devices vary in their tolerance to power supply voltage variations, so most power supplies are controlled by automatic voltage regulators (AVRs).

RAMs need a continuous power supply in order to retain data, and this is sometimes provided by 'floating' the required supply across nickel–cadmium cells. If the supply from the AVR fails for any reason, e.g. loss of mains supply, the voltage to RAM is maintained by the cells with diodes preventing current flowing back into the AVR.

Keyboards and pendants

As seen earlier, most industrial robots are fitted with two devices which enable the operator to input information and instructions directly to the computer control: a keyboard is usually mounted on the control cubicle, and a hand-held pendant is connected to the control by a flexible lead. A typical example was shown in Figure 7.13.

In general the keyboard will present the operator with a button for each letter of the alphabet, laid out alphabetically or in QWERTY form (like a typewriter). Dedicated buttons are often used to avoid tedious button-pressing for commonly used instructions. The keyboard may connect directly to an I/O port of the microprocessor system, or a separate system may be provided to interpret keyboard inputs and to transmit appropriate data for the CPU. From the point of view of safety an emergency stop button is sometimes provided on the keyboard and/or control panel of powerful robots. Unlike most keyboard buttons, the stop switch is hard-wired; when pressed it directly trips the supply to the electric motors or hydraulic power unit of the robot.

A hand-held pendant allows the operator to give instructions to the control while observing the end-effector from any viewpoint. With the exception of the stop button, the controls on the pendant are software operated, i.e. they initiate data signals to input ports of the microprocessor system. Section 7.2 emphasized the desirability of software controls. These allow the CPU to interpret signals from the pendant: individual switches can then cover multiple functions, depending on the mode of operation.

Visual displays

During the teaching process the operator requires information on the state of the system. There are three common methods of providing this: light emitting diodes (LEDs), liquid crystal displays (LCDs), and the cathode ray tube (CRT).

Seven-segment LEDs with green, red or yellow characters and LCDs, familiar in the pocket calculator, are useful for displays in pendants because of their low power requirements.

The CRT is used to provide a VDU in the control cubicle if it is required to present several pieces of information simultaneously. The output to the VDU is memory-mapped, and each character is usually constructed from a matrix of dots 5 wide × 7 high (additional dots separate characters horizontally and vertically).

Inputs and outputs

Input and output connections are required on the robot controller so that grippers and external devices may be operated, and signals from all types of sensors accepted (Coiffet, 1983). These connections may be DC or AC, digital or analogue. When DC is used the standard is 24 V: 110 V is used for AC connections. Outputs are switched or toggled (i.e. pulsed) under control of the program, and inputs may be interrogated by the program. When an input is required the voltage is taken from a common supply in the control cubicle; an external device can then switch the input on by completing the current loop back to the input connection.

A few industrial robots provide for analogue connections. An analogue sensor signal, e.g. from a strain gauge bridge amplifier, may then be accepted directly, and analogue devices, e.g. a pneumatic gripper servo-mechanism for force control, may be driven. Digital to analogue (D/A) and analogue to digital (A/D) converter circuits are used to interface I/O ports with the external connections.

Interfacing

Direct current inputs and outputs are adequate for transmitting individual ON/OFF signals. However, when large amounts of data have to be transferred between the robot control and another computer, for example during the loading of a program written off-line, the standard interfaces or 'bus standards' are necessary. Data is coded on most links in accordance with the American Standard Code for Information Exchange (ASCII), which allocates a number to each letter, digit and function for the purposes of data transmission.

Interfaces may be serial or parallel. A serial link transmits one bit at a time. Although only two wires are required for data transmission, the RS 232C standard, for example, uses a 25-wire connector. The connections are required for the different signals involved in the protocol governing the transmission. For example, to send data a 'request to send' signal is transmitted to the computer on pin 4, and a 'clear to send' signal switched by the computer to pin 5 indicating that data transmission may commence. The number of bits transmitted per second is called the baud rate. 1200, 2400, 4800 and 9600 are common on serial links.

Parallel interfaces can handle higher baud rates because data is transmitted one word at a time, and there is a separate wire for each bit. Examples of parallel links are the IEEE 488 standard, which transmits eight-bit data at up to one million bits/s, using a 24-wire link up to 20 m long, and the S–100 or IEEE 696, operating on a 100-wire link.

A robot computer control

An example of a commercial system is instructive. The Cincinnati Milacron
T3 robot is a hydraulically powered machine, and the jointed-arm manipulator has six degrees of freedom. The elements of the computer control
(Cincinnati, 1980b) of this robot, which is capable of PTP control with linear
interpolation, are shown in Figure 7.15.

All calculations and logical functions are carried out on the processor
board (PRC 32RC) which contains a 16-bit processor system comprising four
Advanced Micro Devices Am 2901B processors. These processors are
four-bit slices using bipolar technology, and are designed for high-speed
controller applications. This board also carries ROM containing the system
software. Data for particular programs can be read from or written to
cassette tape through the interface (cassette I/O), and programs are stored
in 32 K of RAM (MEM) which is protected by the battery back-up unit
(BBU).

The voltage level detection board (VLD) monitors the following DC
supplies from the power supply unit (PSU):

+5 V	logic circuits
+12 V	input, output and servosystem circuitry
−12 V	input, output and servosystem circuitry
−16.75 V	floppy disc (not shown)
+24 V	DC inputs and outputs.

The VLD unit is able to shut down the machine by tripping relays (REL)
should any PSU voltage fail.

The clock unit (CLK) produces 5 kHz sine and cosine signals at 7 V for
the resolvers on the manipulator, and also supplies a 10 MHz signal to the

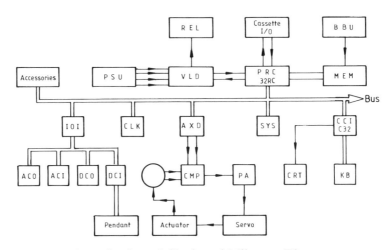

Figure 7.15   Organization of Cincinnati Milacron T3 computer control.

system timing board (SYS), which generates the timing signals for the processor board.

The console control interfaces (CCI) and buffer store (C32) board accepts inputs from the keyboard (KB) using standard ASCII notation and transmits the information to the processor board. This CCI/C32 board also generates a 12 row × 32 column video display from information received from the processor board, and the display is viewed on the cathode-ray tube (CRT).

Sixteen +24 V DC inputs and signals from the teaching pendant are accepted by the DC input board (DCI) through opto-isolators. Similarly sixteen 110 V AC inputs connect to the AC input board (ACI). The DC and AC output boards (DCO and ACO) supply eight outputs each, and these four units are connected to the input/output ports (IOI) by a common 20-line bus.

Most elements of the system can be seen to communicate with the processor board using the common 25-line bus. This is made possible by assigning a device address to each element. Thus the processor is able to set up a communication path to any of the devices using the control and address lines of the bus.

Six axis driver boards (AXDs) are provided for the six actuators. One AXD and its associated components are shown in the diagram. All AXDs are identical, and are distinguished by different device addresses. Each AXD accepts a desired coordinate from the processor and implements the control strategy for that actuator using position and velocity feedback signals from the resolver and tachometer. The AXD generates an error signal and a feedforward signal (which reduces velocity following error). These signals are compensated (CMP) and amplified (PA), driving a servovalve which delivers pressurized oil to the actuator, thus completing the control loop.

7.6 Recapitulation

The digital computer is the only practical means of meeting the demands of a flexible robot control. The different functions which the computer must perform in automatic operation have been detailed. The methods of communicating information to the computer at the teaching or programming stage have been considered. Software to implement these functions has been discussed, working from high-level languages down to the processor's machine language. Finally, the hardware necessary to realize these systems has been detailed, along with an example of a commercial system.

The application of the computer to robot control opens up possibilities for complex sensory information processing, and advanced artificial intelligence systems; these aspects are considered in the next chapter.

Chapter 8

External sensing and intelligence

8.1 Introduction

The last chapter dealt with the problems of monitoring the internal functions of the robot. We now turn to an even more difficult problem—that of relating the robot to its external environment. The earliest robots, and indeed the majority of those in existence today, were relatively stupid: they only did what they were ordered to do. This was often satisfactory, however, especially where repetitive applications of large forces were required or where environments were hostile. These first-generation robots have no awareness of their environment and they will follow their instructions regardless of changes in their surroundings. Indeed, in order to ensure successful operation it is necessary to place them in a carefully structured environment.

The evolution of robots leads to a second generation with the senses of vision, of touch, of hearing and even of smell and taste. These external sensors are particularly important in automated assembly operations, where visual and tactile information is essential. Can you imagine how difficult it would be to pick up a bolt and screw it into a hole if you were blind and had no sense of touch?

Many people recognize that the use of second-generation robots can bring economic rewards. Nevertheless their development and application has been slow, and this has possibly been due to a combination of factors, including the complexity of the technology and the natural reluctance of industrialists to be first in the field (Pugh, 1983). Applications are now growing more rapidly, and many vision packages and a variety of sensors are available on the market.

Having equipped robots with human-like senses, and having given them a variety of other anthropomorphic characteristics, it is natural to wonder if we can go a step further and teach robots to think like humans. This issue is always likely to stir up a passionate debate, for humans have on the whole an innate distaste and fear of machines that are capable of apparently intelligent behaviour. Nevertheless, the third generation of robots will have artificial intelligence: they will be able to act in an apparently rational manner; they will be able to understand and respond to natural language; they will be able to solve complicated problems in areas that would normally require a human expert. But before discussing artificial intelligence in any depth it is necessary to look at robotic sensors.

8.2 Touch and tactile sensing

It is necessary at the outset to distinguish between touch and tactile sensing. Following L. D. Harmon, we note that tactile sensing involves the continuous measurement of forces in an array (Stauffer, 1983; Harmon, 1983). It uses skin-like properties. Touch, on the other hand, refers to simple contact for force sensing at one or just a few points.

Figure 8.1 shows three examples of touch sensing. The first (Figure 8.1(a)) is typical of many applications where the robot merely needs to be informed if it has grasped an object or if it is in the correct position to initiate the grasping action. As shown, a simple microswitch is often satisfactory, although if items are delicate it may be necessary to use a whisker-operated switch or indeed to revert to non-contact sensing (see below).

Robot welding systems require seam tracking systems that enable the robot to detect and to correct deviations from the desired weld path. We

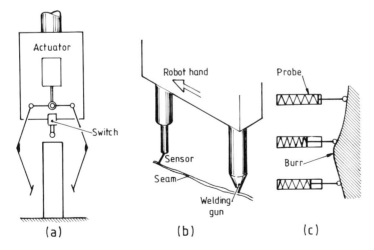

Figure 8.1 Touch sensing: (a) use of a microswitch (b) a probe for guiding a welding gun (c) detecting burrs.

shall see later how robot vision has been used to solve this problem, but here we refer to a solution that relies upon touch sensing (Figure 8.1(b)). The sensor consists of a needle-like probe which is situated about 3 cm ahead of the welding gun (Presern *et al.*, 1981). The needle has two degrees of freedom with a maximum displacement of ±10 mm. Optical sensors are used to convert the needle's displacement to a digital signal which is used to guide the welding gun. A resolution of 0.05 cm is claimed. This system avoids the need for two passes along the seam—one to teach and one to weld. Hence the operation is carried out more quickly and no computer storage is required for the seam information.

Another example of touch sensing is shown in Figure 8.1(c). Here miniature potentiometers are connected to spring-loaded probes for determining the rough outline of a component. This method has been extended (Abele, 1981) to determine the size of burrs during the fettling of castings, the information being used for calculating the ideal path for the grinder.

Touch sensing gives crude information about contact with an object. Tactile sensing, on the other hand, allows the distribution of contact forces to be determined, and this can be used to calculate the magnitude and direction of the gross contact force. In addition, and perhaps of greater significance, a knowledge of the distribution of the contact forces can be used to identify the object, and this is one important aspect of artificial intelligence. Practical systems use an artificial skin consisting of arrays of force sensing elements, or tactels, placed on the fingers of an end effector. Several tactile sensors are available commercially (Allan, 1983).

Harmon (1983) has carried out a comprehensive survey of tactile sensors. He lists the following desirable characteristics:

- The sensor surface should be compliant and durable.
- Spatial resolution should be around 1–2 mm.
- A range of 50–200 tactels would be acceptable.
- Sensors must be stable, monotonic and repeatable with low hysteresis.
- Elements should have response times of 1–10 ms.
- Dynamic range should be 1000:1 with a threshold sensitivity of about 0.01 N/m^2.

One of the earliest tactile elements (Larcombe, 1981) relied on the change of resistance of a felt form of carbon fibre. Increasing pressure (Figure 8.2(a)) forces the individual fibres into closer contact thereby reducing the electrical resistance between the electrodes. Since the felt is flexible it can be tailored to the gripper surface.

Another promising technique (Purbrick, 1981) uses the variation of resistance at the interface between orthogonal silicone rubber cords (Figure 8.2(b)). In the unstressed state the area of contact between cords is small and resistance between electrodes is therefore high. The application of pressure increases the area of contact and reduces the resistance. A 16 × 16 matrix has operated successfully, the 256 individual sensors or tactels being scanned automatically by combined analogue and digital circuits.

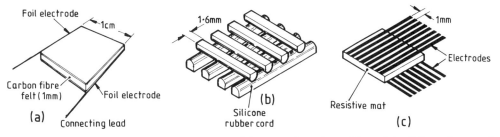

Figure 8.2 Tactile sensing elements: (a) use of carbon fibre felt (b) orthogonal silicone rubber cords (c) an 8×8 array using a piezoresistive material.

Some researchers have rejected rubber materials because of suspected fragility, hysteresis and non-homogeneity, the last factor becoming of increasing importance as tactel size is reduced. This has led to the development (Robertson and Walkden, 1983) of sensors which use a specially developed piezoresistive material made from a fine yarn impregnated with a conductive medium. As with carbon felt and rubber, the resistance across the device reduces with pressure. Figure 8.2(c) shows the layout of an 8×8 array with electrodes set at a pitch of 1 mm. A 16×16 array is under development with the capability of providing pressure images for object recognition. It is being used in conjunction with a parallel-link robot similar to that illustrated in Figure 2.10.

Finally under the heading of piezoresistive materials, it is interesting to note the development of a solid-state VLSI tactile sensing chip (Raibert and Tanner, 1982). Forces are transduced using conductive plastic, but the outstanding feature is an active substrate that handles the necessary computing and communications. An array of processors in the sensor performs filtering and convolution operations (see Section 8.6) on the tactile image. Research has been carried out on a 6×3 array with 1 mm square tactels, but plans for a 30×30 array are in hand.

The tactile sensors so far discussed have relied on change of resistance with pressure. We now turn to an examination of sensors which use magnetic effects. Figure 8.3(a) illustrates a tactile element constructed from a magnetorestrictive nickel-based metal (Chechinski and Agrawal, 1983). A magnetic material is magnetorestrictive if it shows a change in its B field when subjected to external forces. The sensor, known as a transformer–pressductor, is 7 mm high, 5 mm wide and 3.5 mm thick. The primary and secondary windings are orthogonal and have 10 turns each. Application of force causes a change in the induction vector and this in turn changes the secondary voltage. Tests were carried out up to a maximum of about 1 N and the results demonstrated linearity between secondary volts and force, with little hysteresis.

A more recent development (Vranish, 1984) uses a different magnetic effect, called magnetoresistance. Magnetoresistive materials such as Permalloy (81/19 NiFe) show changes in resistance when under the influence of varying magnetic fields. This effect is put to good use in the sensor of Figure

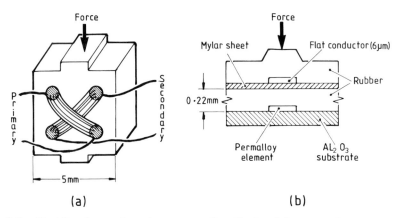

Figure 8.3 Tactile elements using magnetic effects: (a) a transformer-pressductor (b) use of a magnetoresistive material.

8.3(b). A skin has been constructed of a thin-film magnetoresistive array with sensor elements covered by a sheet of rubber and a row of flat wires etched on a mylar film. Tactels are spaced at 2.5 mm intervals, an 8×8 array taking up an area of 25 mm square. Thickness can be as little as 2.5 mm. The applied force reduces the distance between the conductor and the permalloy element and the resultant change in the magnetic field at the element causes its resistance to change. Pressure thresholds can be as low as $30 \, N/m^2$ or as high as $2 \, MN/m^2$, depending on the thickness of the top layer of rubber.

There are also several tactile sensors that employ optical effects. The sensor of Figure 8.4(a) uses a mechanical deflector and an electro-optical transducer (Rebman and Morris, 1983). Movements of the pin, which is integral with the compliant surface material, affect the amount of light passing between an LED and a phototransistor. The resultant current is a

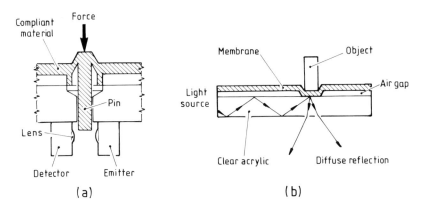

Figure 8.4 Tactile elements using optical effects: (a) mechanical interruptions of a light beam (b) use of diffuse reflections.

measure of the applied force. An 8×8 array measuring 81 mm square and measuring forces up to 4 N is available commercially.

Another way of employing optical effects in a tactile sensor is illustrated in Figure 8.4(b). The sensor uses the properties of reflection between objects of different refractive index. The transducer (Mott *et al.*, 1984) consists of a light source and a clear acrylic plate separated from a compliant membrane by an air gap. Since the refractive index of the acrylic is greater than that for air, a sheet of light directed along the edge of the plate will be internally reflected and will pass out at the right-hand end. However, if pressure is applied to the membrane it will contact the acrylic, and diffuse reflections will occur because the refractive index of the membrane is greater than that of the acrylic. Light will escape sideways, the intensity of the bright patch being proportional to the applied pressure. This light can be detected by a camera (Mott *et al.*, 1984) or by an array of phototransistors (Tanil *et al.*, 1984). It is claimed (Mott *et al.*, 1984) that the resolution of a tactile sensor of this form can match that of high-resolution vision sensors (see below), typically 256×256 tactels on a 2 cm square pad.

Slip sensing

When a robot or a prosthetic hand has to lift fragile objects it is important that the minimum gripping force be used. The minimum gripping force is that which just provides enough friction to stop the object from slipping. There are several ways of detecting slip. Microphones built into the fingers of prosthetic hands have been used to detect slip by the sound it generates (Swain, 1982). Another method uses a roller built into a robot gripper and bearing against the part to be lifted (Masuda and Hasegawa, 1981). Slippage causes rotation of the roller and this is detected by photoelectric sensors. It should be noted that all of the above-mentioned tactile sensors have the capability of sensing slip as well, for if slip occurs the tactile image changes and this could be used to indicate that slip has taken place.

8.3   Measurement of gross forces and torques

Tactile sensing can provide information about the position of an object, about slippage and, if the resolution is high enough, about the shape of the object. Tactile sensing is basically concerned with the reaction forces between the gripper and the object. Force sensing, on the other hand, provides information about the reaction forces between the object and the external environment. Its major area of application is in mechanical assembly. For example, when placing a peg in a hole the robot control system uses force information to adjust the position and orientation of the peg until reaction forces are minimized. This is known as active compliance, in contrast to the passive remote centre compliance referred to in Chapter 2. Force-reflecting tele-operators, such as those described later in Chapter 10, provide another area of application.

In order to achieve active control it is necessary to know the generalized force vector, i.e. the unknown force between the end-effector and the environment is resolved into three separate forces, F_x, F_y and F_z along orthogonal axes, and three separate moments M_x, M_y and M_z around those axes. There are several ways of doing this.

By measuring forces and torques at each machine coordinate

If forces and torques can be measured at each actuator then it is possible to deduce the generalized force at the gripper (Nevins *et al.*, 1974).

Let the vector dX be a small motion of the gripper, expressed in world coordinates, the vector $d\Theta$ the corresponding machine coordinates, and J the Jacobian. Then if the arm is sufficiently rigid we can write

$$dX = J\, d\Theta \qquad (8.1)$$

If these displacements are the result of a generalized force F applied to the hand, and if there are no energy losses, then eqn (5.1) allows us to write

$$F^t\, dX = F^t\, d\Theta = T^t\, d\Theta + W^t\, d\Theta \qquad (8.2)$$

where T is the vector of joint forces and torques, and W is the vector of gravity-induced couples. Hence

$$F^t = T^t J^{-1} + W^t J^{-1} \qquad (8.3)$$

This is the desired result, showing how the generalized force can be found from the joint forces/torques and the gravity couples.

A major shortcoming of this method is the difficulty of measuring the joint torques with sufficient accuracy, due to their contamination by friction effects.

By using pedestal force sensors

The forces of interaction between a robot-held object and earth can be determined from measurements taken on a force pedestal attached to earth. Such pedestals have long been used for the measurement of unknown forces and moments in windtunnel models, machine tools and rocket engines. Figure 8.5(a) shows a six-component thrust stand used in testing rocket engines (Doebelin, 1983). Six load cells are used, three measuring forces F_4, F_5 and F_6 along the sides of an equilateral triangle in the XY-plane, and three measuring forces F_1, F_2 and F_3 in the vertical Z-direction. The test engine is attached to the common centre of both triangles. The components of the

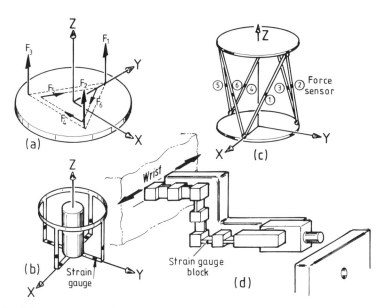

Figure 8.5 Measurement of gross force/torque: (a) a six-component thrust stand (b) a pedestal force sensor (c) a force sensor for wrist mounting (d) a sensor mounted directly on the gripper.

generalized force vector are

$$
\left.
\begin{aligned}
F_x &= 0.5(F_5 + F_6) - F_4 \\
F_y &= (\sqrt{3}/2)(F_5 - F_6) \\
F_z &= F_1 + F_2 + F_3 \\
M_x &= (0.5l/\sqrt{3})(2F_1 - F_2 - F_3) \\
M_y &= 0.5l(F_3 - F_2) \\
M_z &= -(0.5L/\sqrt{3})(F_4 + F_5 + F_6)
\end{aligned}
\right\}
\tag{8.4}
$$

where l and L are respectively the sides of the large and small equilateral triangles.

The device of Figure 8.5(a) was first used in 1960. Since then several other pedestal force sensors have been developed, some especially for robotic assembly applications (Nevins *et al.*, 1974). One example, used in conjunction with a pick-and-place robot, is shown in Figure 8.5(b). The sensor consists of a framework of eight stainless steel beams, each carrying a semiconductor strain gauge bridge (Kasai *et al.*, 1981). Knowing the strain in each bridge it is possible to determine the generalized force. A torque sensitivity of 2×10^{-3} N m and a force sensitivity of 2×10^{-2} N are claimed. When a component is being inserted by the robot, the pedestal measures the forces and torques and the controller accordingly tilts or moves the table laterally to improve alignment.

By mounting a generalized force sensor on the robot

When a robot is designed for general assembly work there is a lot to be gained by making the generalized force sensor an integral part of the robot. The device shown in Figure 8.5(c) is fitted between the robot's wrist and end-effector, the end-effector being attached to the lower plate and the wrist to the upper (Gaillet and Reboulet, 1983). The geometry is similar to that of the parallel actuator robot of Figure 2.10, but here the six links are fixed in length and each has a force sensor attached.

If the plates have equal radii R, the vertical distance between plates is h and the length of a link is L, then the relationship between the generalized force and the measured forces F_1 to F_6 is

$$\begin{bmatrix} F_x \\ F_y \\ F_z \\ M_x \\ M_y \\ M_z \end{bmatrix} = \begin{bmatrix} -a & 2a & -a & -a & 2a & -a \\ 2\sqrt{3}a & 0 & -\sqrt{3}a & \sqrt{3}a & 0 & -\sqrt{3}a \\ b & b & b & b & b & b \\ 0 & -\sqrt{3}ha & -\sqrt{3}ha & \sqrt{3}ha & \sqrt{3}ha & 0 \\ 2ha & -ha & -ha & -ha & -ha & 2ha \\ -\sqrt{3}aR & \sqrt{3}aR & -\sqrt{3}aR & \sqrt{2}aR & -\sqrt{3}aR & \sqrt{2}aR \end{bmatrix} \cdot \begin{bmatrix} F_1 \\ F_2 \\ F_3 \\ F_4 \\ F_5 \\ F_6 \end{bmatrix}$$

$$(8.5)$$

where $a = R/2L$ and $b = h/L$.

This is much more complicated than the transformations required for the triangular pedestal force sensor described in eqn (8.4). In that case determination of a component of the generalized force vector required a calculation involving at most three of the measured forces. Here four of the vector's components involve all six measured forces in their calculation. Whilst recognizing this strong cross-coupling, Gaillet and Reboulet (1983) argue that this shortcoming is more than offset by the sensor's compactness, by its ability to account for temperature effects, by its easy tuning and by the absence of the numerous correction terms that arise in other sensors.

Installing a force sensor between the wrist and the end-effector may not be entirely satisfactory if the end-effector is particularly massive: the weight of the end-effector could take up most of the sensor's dynamic range. The solution is then to mount the sensor directly on the gripper, as in Figure 8.5(d). Here six interchangeable modular sensors are used (Wang and Will, 1978), each consisting of a flexible beam with a strain gauge attached. The orthogonality of the structure results in a considerable reduction in the size and complexity of the transformation matrix.

8.4   Proximity detection using non-contact sensing

Having dealt with those aspects of robotic sensing that relate to the human sense of touch, we now turn our attention to non-contact sensing. We shall concentrate initially upon proximity sensors whose primary function is to determine whether an object or part of an object is within a specified distance from the robot's end-effector. The simplest touch sensors were earlier shown to be used for determining if an object has been grasped or if it were in the right position to be grasped. Non-contact sensors have two important advantages over touch sensors: they will not damage the sensed object, and, not being subject to repeated contacts, they have a longer life.

Many physical effects are employed in non-contact proximity sensors—pneumatic, acoustic, magnetic, electrical and optical being the most common.

Pneumatic proximity switches

Pneumatic sensors are of two types: back-pressure sensors and interruptible jets (McCloy and Martin, 1983). The former, which include air gauges, have been in use for many years. Their principle of operation is illustrated in Figure 8.6(a) which shows a tube-shaped probe in which a fixed supply pressure P_s is connected by a fixed orifice to a control chamber at pressure P_c. The magnitude of P_c depends on the size of the variable orifice formed by the end of the tube and the object whose distance or presence is to be detected. The nearer the object, the less escape of air and the higher the pressure P_c. When the escape area is less than that of the outlet orifice, i.e. when $x < 0.25D$ the pressure P_c varies proportionally with the distance x (Figure 8.6(b)). So, as well as being able to detect the presence of an object, the sensor can be used to measure distances within this region. The problems of air loss and sensitivity to stray currents usually restrict the

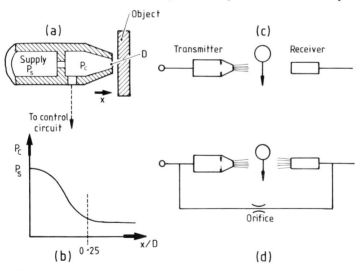

Figure 8.6 Pneumatic proximity detectors: (a) back-pressure sensor (flapper nozzle) (b) variation of pressure with distance to object (c) an interruptible jet (d) interruptible jet with biased receiver pressure.

diameter D to a maximum of about 4 mm, so the dynamic range is restricted to about 1 mm in most applications (Belforte *et al.*, 1981).

The interruptible jet (Figure 8.6(c)) is another form of pneumatic proximity sensor. It is the pneumatic equivalent of the photo-emitter and receiver (see below), and is thus sometimes referred to as the pneumatic eye. Figure 8.6(c) illustrates the principle of operation. A low-pressure source transmits a jet of air across a gap to a receiving orifice where some pressure is recovered. If an object interrupts this jet the receiver pressure falls and the change in signal can be used to indicate the object's presence. Typical jets have transmitter diameters around 3 mm, receiver diameters around 6 mm, gaps of 25–50 mm and supply pressures from 0.07–0.2 bar (McCloy and Martin, 1983). In dusty environments, such as flour mills or cement works, the possibility of the receiver clogging can be reduced by pressurizing the receiver, as shown in Figure 8.6(d).

Although it would be possible to detect objects by mounting an interruptible jet sensor on the jaws of the robot gripper, the sensitivity to stray draughts could cause problems during robot motion. Parts inspection and identification are more promising areas of application, although even here the power lost in escaping air cannot be ignored.

Acoustic proximity detectors

Echo effects can be used for proximity detection, although in robotics the main area of application is likely to be in range finding (see Section 8.5). The main element in this type of sensor is an ultrasonic generator, often based on the resonance of a piezocrystal. The requirements of a narrow beam makes the use of ultrasound (>20 kHz) necessary. The acoustic sensor relies on the reflection of this transmitted beam from a target object, the detection of a reflected acoustic signal indicating the presence of the object. In some sensors the same transducer is used for both transmission and detection.

Acoustic proximity sensors can detect objects within a range roughly 5–100 cm. It has been argued (Luo *et al.*, 1983) that their relative bulk precludes the possibility of their installation at the robot gripper.

Magnetic/electric proximity detectors

The simplest form of non-contact magnetic proximity sensor is the dry reed switch. Two thin strips or reeds of magnetic material are housed in a glass envelope with external connections to an electrical circuit. Normally the reeds are apart and the circuit is open, but if the sensor enters the vicinity of a magnetic field the reeds attract each other and make electrical contact (Figure 8.7(a)).

Another type of non-contact sensor, the inductive proximity switch, relies on currents induced in the target object. Figure 8.7(b) shows the basic elements. A coil of wire is moulded into the sensing face and forms part of a high frequency oscillator circuit. The oscillating magnetic field which exists

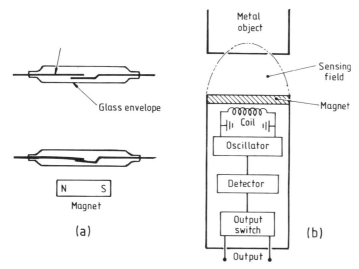

Figure 8.7 Magnetic/electrical proximity detection: (a) the dry reed switch, open and made (b) an inductive proximity switch.

around this coil, causes eddy currents to be induced in any conducting target that enters the field, and the resultant effect on the amplitude of the oscillations can be used to operate an output switch (Kay, 1983; Luo *et al.*, 1983). This sensor can detect metallic objects approaching from any direction. Sensitivity is highest for ferrous targets, sensing distances being approximately halved for the non-ferrous metals. Where targets are non-conducting it has been found sufficient to attach a strip of adhesive-backed aluminium foil to the surface. Most switches are totally encapsulated and cylindrical in shape, with diameters ranging from 2 to 60 mm. Sensing ranges usually lie between 0.25 and 40 mm.

Changes in capacitance between an object and a sensing head are used in capacitive proximity sensors. These are very similar to the inductive type but with one important difference: the active part of the capacitive sensor is a layer of metal plates mounted behind the sensing face. The plates form part of the capacitive element of an oscillator. When an object is within the sensor's range of sensitivity, the changing geometry and/or dielectric characteristics cause the capacitance to change, disturbing the oscillator and providing a switching signal. Whilst the inductive proximity sensor is only affected by conductive targets, the capacitive sensor has the advantage of reacting to most materials, including plastics, wood and glass. The size and dynamic range of the capacitive sensors are very similar to their inductive counterparts.

Optical proximity detectors

Optical methods for non-contact monitoring and measurement offer advantages over other methods, particularly their inherent safety with respect

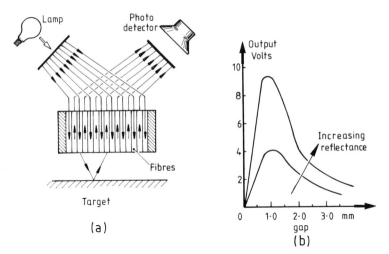

Figure 8.8   The Fotonic Sensor: (a) general arrangement (b) output as a function of distance to target and reflectance of surface.

to electrical or fire hazards. The simplest devices use the amount of light reflected from a target as a measure of distance. One of the earliest commercial sensors, the Fotonic sensor (Menadier *et al.*, 1967), used fibre optics to measure displacements in a range from micrometres to millimetres (Figure 8.8(a)). Light is transmitted onto a target through a bundle of optical fibres, and the target reflects some of this light into receiving fibres which transmit it to a photodetector. Probes vary from 0.5 to 7.5 mm in diameter and consist of a bundle of optical fibres each about a tenth of a

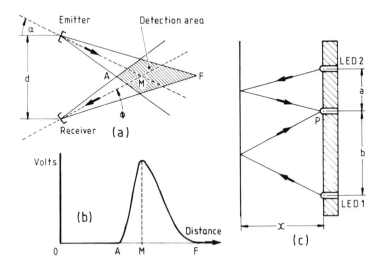

Figure 8.9   Optical proximity detectors: (a) infrared light is reflected from a target in the detection area (b) receiver output as a function of target distance (c) an optical detector that uses time of flight.

millimetre in diameter. A typical variation of output voltage with target distance is illustrated in Figure 8.8(b). There are two useful measurement ranges, front slope and back slope, but the front slope is usually preferred since it is more sensitive and more linear.

Infrared LEDs are now commonly used for proximity detection. Their use reduces the possibility of interference from ambient lighting. A commonly used sensor uses an LED as an emitter and a phototransistor as a receiver (Figure 8.9(a)). Infrared light (wavelength 900 nm) is beamed onto a target and the intensity of the reflected light, as seen by the photo-transistor, is a function of the distance to the target (Espian and Catros, 1980). The beams are coplanar, they are tilted symmetrically by angle α, and the emitter and receiver are separated by distance d. In practice the sensor produces a narrow cigar-shaped sensitive volume permanently focused a few centimetres in front of the sensor. When a reflective object enters this volume the resultant voltage output at the phototransistor is a non-linear function of the target distance. Figure 8.9(b) shows a typical output variation when the target plane is orthogonal to the sensor plane. The distances OA, OM and OF are given by

$$\left. \begin{array}{l} OA = (d/2)\tan(\alpha + \phi) \\ OM = (d/2)\tan \alpha \\ OF = (d/2)\tan(\alpha - \phi) \end{array} \right\} \tag{8.6}$$

where ϕ is the beam spread, determined by the optics. It can be seen that precision and range depend on the values of d and α. For example, a large value of d and a small α result in a sensor with wide range and small precision.

Figure 8.9(b) shows that the output is the same at two different distances. This can be confusing, but the problem can be avoided by recessing the sensor inwards so that only the back slope of the characteristic is used (Bejczy, 1980). Perhaps a more serious problem is presented by the effects of surface orientation and reflectivity, but in spite of these difficult there have been several successful attempts to integrate these sensors in the structure of an end effector. Espian and Catros (1980) describe their u in a 'smart hand' in a tele-operated aid for a quadriplegic (see Chapter 9 and Bejczy (1980) illustrates their application in a space tele-operator.

The two optical devices described above rely on variations in th intensity of reflected light. Their performance is therefore affected by variation in the reflectance of detected objects. In cases where such variations are likely to be large, the use of 'time of flight' information offers a better solution. Knowing the velocity of light c, and knowing the time t taken by a beam of light to pass from a transmitter to a receiver, then it is an easy matter to determine the length L of the path from $L = ct$. This time lag, or phase shift, which is the same thing in another guise, can be used to measure distance. Trounov (1984) describes a distance sensor which uses the phase shift between two beams for distance measurement. Two LEDS are placed asymmetrically with respect to a phototransistor P (Figure 8.9(c)). LED 1 is modulated by $\sin \omega t$ and LED 2 by $\cos \omega t$. The phase shift ψ between the beams arriving at the phototransistor is given by

$$\psi = (\pi/4) - (\omega/2c)\{\sqrt{4x^2 + a^2} - \sqrt{4x^2 + b^2}\} \qquad (8.7)$$

It is claimed that the device can detect within a range of 5 to 70 mm, to an accuracy of ± 1 mm. The dimensions a and b are respectively 30 mm and 10 mm, and the total processing time is 5.3 ms.

8.5   Range-finding

Non-contact proximity sensors detect objects or measure distances to objects within a restricted range, typically millimetres to centimetres. When distances are greater, metres or kilometres, it is necessary to employ range-sensors. These rely on the transmission of ultrasonic beams or light beams.

Ultrasonic range-sensors

It was indicated earlier in Section 8.4 that ultrasonic sensors can be used for proximity detection in the range 5–100 cm, but their main area of application is in range-finding. Acoustic range-sensors deduce target distances from measurement of the delay in the echo of an ultrasonic pulse. They have been used in the automatic focusing system of a popular camera (Polaroid, 1980). The simpler acoustic range-finders use a single frequency system, but since a camera range-finder must deal with targets of variable and unpredictable geometry, it is necessary to employ four different frequencies to ensure a reliable echo. The transmitted signal consists of 8 cycles of 60 kHz, 8 cycles of 57 kHz, 16 cycles of 53 kHz and 24 cycles of 50 kHz—a total of 56 cycles in about 1 ms. An electrostatic transducer is used alternately as transmitter and receiver. The system was designed for measurements in the range 25 cm to 10 m with a resolution of 3 cm. It has been modified for use in a sensor-controlled gripper (Dillman, 1982). The minimal measurable distance was limited to 18 cm because of time taken for transmission and the time taken to switch the transducer from transmitter mode to receiver mode.

Range-finding is an essential requirement of mobile robots. In one example (Banzil *et al.*, 1981), the position of a roving robot with respect to obstacles, corners, etc., is determined by an array of 14 onboard ultrasonic emitter/receiver modules. Each unit is stimulated in turn and emits four cycles at 36 kHz, and distance is obtained by counting the number of impulses on a 1.536 MHz clock. Distances around 2 cm can be measured with an accuracy of 0.5 cm. A commercially available educational robot uses a single transducer, generating eight cycles of 32 kHz at 0.5 s intervals (Heathkit, 1980).

Optical range-sensors

The accuracy of acoustic devices is affected by variations in the speed of sound due to atmospheric conditions. Better accuracy is obtained using optical sensors. The two most commonly used techniques are (a) time of flight or LIDAR (light detection and ranging) and (b) optical triangulation.

LIDAR systems can be further subdivided into pulsed and continuous-wave systems. In the former, the time for a light pulse to travel from the source to object and back is measured. This technique is useful for the larger ranges, say >10 m. It cannot be used for ranges less than about 2 m, since with minimum practical pulse widths of about 5 ns the pulse itself is spread out in space over 1.5 m (the speed of light is 3×10^8 m/s). It would also be extremely difficult to achieve high resolution—for example, a resolution of 6 mm would require the detection of time differences as small as 20 ps. These limitations restrict the use of pulsed LIDAR as far as robots are concerned. The most promising area appears to be the guidance and control of mobile robots, where distances can be relatively large (5–10 m) and resolution is not so demanding (Lewis and Johnson, 1977).

Continuous-wave modulated LIDAR offers more promise. In this case range data are determined from the phase shifts between the transmitted and received signals. The system described by Nitzen *et al.* (1977) uses a modulation frequency of 9 MHz, and can measure ranges from 1–5 m with a resolution of 1 cm (equivalent to 0.108° phase shift). A major problem is the large dynamic range ($\cong$100 dB) caused by variations in target reflectance with distance.

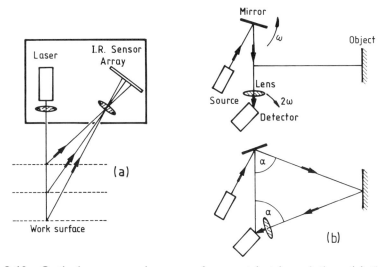

Figure 8.10 Optical range sensing: use of geometric triangulation: (a) the position of the reflected ray on the sensor array is a measure of distance (b) use of a rotating mirror and lens.

Geometric triangulation is another technique commonly used in optical range finding. Two examples are shown in Figure 8.10. In the first, infrared light from a solid-state laser is reflected from an object surface and passes through a lens onto an array of infrared sensors. The position of the point of light on the array is a measure of the range. A commercially available device (Edling and Porsander, 1984), used for following seams in arc-welding applications, has a mass of 0.65 kg, a resolution of 0.06 mm and a range of 32 mm about a mean distance of 170 mm.

Another example of triangulation (Figure 8.10(b)) comes in a compact $7 \times 7 \times 8$ cm package (Nimrod *et al.*, 1982). As shown, a transmitting mirror, rotating at a constant rate ω, sweeps a laser beam in a plane at rate 2ω. The distance between the axis of the mirror and the photodetector is the baseline B. A focusing lens rotates around the photodetector at -2ω and is synchronized with the mirror so that the angle α between the baseline and the beam from the mirror is always equal to the angle between the baseline and the optical axis of the lens. When a target is placed somewhere along the measurement axis, two signals are generated by the photodetector for each rotation of the mirror. The first occurs when $\alpha = 0$ and the second when the mirror beam illuminates the target at its intersection with the measurement axis. The range R is given by

$$R = 0.5B \tan \alpha \tag{8.8}$$

and the angle α is given by

$$\alpha = 4\pi T_a / T_c \tag{8.9}$$

where T_a is the time lag between the first and second signals, and T_c is the mirror scan cycle time. Resolution is affected by range, baseline and variation in scanning speed. Precision varies from 0.07% of range at 1 m to 0.1% at 30 mm.

8.6 Machine vision

From optical range-sensing it is only a short step to machine vision and thence to artificial intelligence. In range-sensing the determination of the object's distance is the chief objective, but machine vision, although often still requiring range information, is more concerned with using that information to determine not only the object's distance, but its position, its orientation, its shape and indeed its identity. Rosen (1979) lists the major functions for machine vision in robotics as:

- The recognition of workpieces/assemblies and/or recognition of the stable state where necessary.
- The determination of the position and orientation of workpieces/assemblies relative to a prescribed set of coordinates axes.
- The extraction and location of salient features of a workpiece/assembly to establish a spatial reference for visual servoing.

- In-process inspection: verification that a process has been, or is being, satisfactorily completed.

Image acquisition

The first stage in any vision system is that of image acquisition, which requires a camera and a means of illumination. There are two basic types of illumination, reflected illumination and through illumination. Reflected illumination (PERA, 1982) yields details of the surface features of an object, but this requires carefully controlled lighting and powerful software to cope with the multi-grey-level image. Through illumination, or backlighting, produces binary black and white images and, whilst these require much less data processing than grey-level images, the resulting information is restricted to the contour of the object. The choice of technique depends on the specific application.

The camera is basically a transducer which converts the luminance of an array of image points into an array of electrical signals. In thermionic tube cameras, such as Vidicon, the electrical resistance of individual points on a photoconductive screen varies with the intensity of the light falling on the points. The screen is scanned row by row, left to right, from top to bottom, the resulting output being known as the video signal.

Vidicon camera technology, although cheaper, is gradually giving way to solid-state technology. The majority of solid-state cameras use charge-coupled devices (CCD), and in comparison to their Vidicon counterparts they are less bulky and more robust, have better geometric stability, less drift and fewer problems with blooming. A typical CCD solid camera uses a 256×256 photodiode array, giving a total of 65 536 picture elements or pixels (Boucharlat *et al.*, 1984). Each pixel is $29\,\mu$m square and the total area of the photosensitive zone is only 7.4 mm square. Solid-state linear arrays with up to 2048 pixels are also available, but for completion of an image it is necessary to arrange for relative motion of the object across the line of view of such a camera.

Image processing

The amount of data captured by a camera can be quite staggering. The video signal is first of all subjected to an analogue-to-digital conversion either to a simple binary black and white picture, or to a picture with as many as 256 grey levels (eight bits). When processing an image it is necessary to store at least one frame, and in a 512×512 pixel image with eight bits of grey level information such a frame would contain 2 097 152 bits. In microcomputers such as the Pet or Apple, where eight-bit words are used, this will require 262 144 words—which the micro cannot handle. In 32-bit minicomputers such as the VAX 11/750 this is reduced to 65 536 words, which is a very heavy demand. Obviously this can be reduced by cutting down on resolution and/or grey level.

The processing and interpretation of these data falls broadly into two tasks (PERA, 1982):

- Routine processes concerned with the preparation of data for subsequent analysis. These are mostly carried out by dedicated hardware, typical tasks being the removal of noise and increase of contrast.
- Analysis. This usually relies on software and includes such tasks as edge detection, measurement of perimeter and areas, determining position and orientation, identification.

Let us consider a typical analytical task, that of edge detection. Edges are detected by searching for sharp changes in intensity gradient. If there were no noise this would be a relatively simple task, but with noise and surface imperfections it is necessary to seek a compromise between maximizing the detection of real edges and minimizing the detection of noise edges. This is achieved by using a mathematical operator to enhance the edges and then applying a threshold to distinguish the edges from the rest of the image. Many commonly used operators approximate the intensity gradient ∇G, where

$$\nabla G = \frac{\partial G}{\partial x} + \frac{\partial G}{\partial y} = G_x + G_y \qquad (8.10)$$

$$|\nabla G| = \sqrt{G_x^2 + G_y^2} \qquad (8.11)$$

and G is the intensity at point (x, y). A digital approximation for ∇G can be obtained with the help of Figure 8.11(a) which shows a pixel (i, j) and its eight neighbours a_0 to a_7. The partial derivative in the x-direction can be approximated by

$$G_x = (A_2 + cA_3 + A_4) - (A_0 + cA_7 + A_6) \qquad (8.12)$$

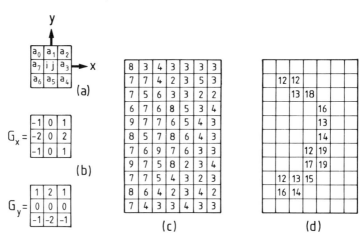

Figure 8.11 Edge detection: (a) a pixel (i, j) and its eight neighbours (b) Sobel operator masks (c) an intensity matrix (d) edge detected by use of the Sobel operator.

and in the y-direction

$$G_y = (A_0 + cA_1 + A_2) - (A_6 + cA_5 + A_4) \qquad (8.13)$$

where A_n is the intensity at pixel a_n and c is constant. When $c = 1$ we have the Prewitt operator, and when $c = 2$ the Sobel operator (Nevatia, 1982). The direction of an edge is given by $\theta = \tan^{-1}(G_x/G_y)$.

It is often easier to handle these operators when they are written in 'mask' form, i.e. as configurations of values which will be multiplied by the corresponding intensities at the pixels, and then added or subtracted as the sign of the coefficient in the mask suggests. Figure 8.11(b) shows the masks for the Sobel operator described in eqns (8.12) and (8.13), with $c = 2$. The results of using these masks or filters to find an edge in the intensity matrix of Figure 8.11(c) are shown in Figure 8.11(d). A threshold of 12 units was applied. This process, called convolution, is slow when masks are large, and it has been found that results are obtained more quickly when convolution is performed in the frequency domain. This requires multiplication of the discrete fast Fourier transforms of both mask and image.

Davies (1984a) describes how the Sobel operator is used in a vision system for examining food products, particularly biscuits. It emphasizes that thresholding the enhanced image is less prone to error than thresholding the original. However, as shown in Figure 8.11(d), edges appear thick in some places and may peter out in others. There are methods for correcting these deficiencies, but edge thinning in particular requires a considerable computational effort.

Object recognition

Once the boundary of the object has been found, a boundary-tracking procedure may be employed to pass the information to a shape-analysis algorithm. One approach is to compute the centroid of the image and then to plot the boundary on an (R, θ) graph as shown in Figure 8.12. This graph

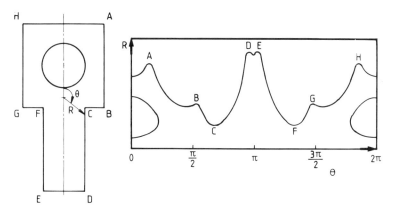

Figure 8.12 Object recognition: the (R, θ) graph is unique to a particular component.

has a period of 2π, and for a known object will merely be displaced by some fixed angle α from the (R, θ) graph of an ideal form of the same object in standard orientation Davies (1984). This ideal form, or template, would be kept in memory, and it is a relatively easy task to shift the template data with respect to the measured data until a match occurs. This gives the orientation α of the component. If the match is less than satisfactory then the object may be discarded as being less than perfect, or indeed of the wrong sort.

Angular matching is a limited form of the more general problem of *template matching*. If a computer vision system is to be used to determine the location and orientation of an object, and to recognize it, then an enormous amount of computer power may be required. For example, if an object of size about 30×30 pixels is to be located in an image of 256×256 pixels then the task of searching for a match by moving the template around the image could involve as many as $256^2 \times 30^2$ ($\cong 60$ million) basic operations (Davies, 1984). Assuming, optimistically, that the computational cycle takes $1\,\mu s$, then the time for matching could be as long as 1 min for this simple case. Such a long processing time would exclude such a vision system from on-line applications in robotic handling.

Fortunately there are several ways of speeding the operation up. In one, the system is programmed to seek out salient features such as corners or holes, and when such a feature has been located it is a relatively small task to search for and locate a secondary feature at a known distance from it. The identification of both of these features confirms the identity of the object and gives information about its location and orientation.

Another method of reducing the computational load performs calculations on local areas only, and global relations are derived from combinations of these local calculations, in a presumed analogy with the human visual system (Nevatia, 1982). Logical n-tuple methods (Aleksander, 1983) fall into this category, an n-tuple being a group of pixels which is processed as a single entity. For a simple task it has been demonstrated that this method requires 1/25th of the storage required for template matching (Aleksander, 1983).

The above method attempts to learn from the parallel processing used in the human visual system. The advantages of parallel processing are well recognized, and are now being applied to vision systems. In serial processing the processor scans the image sequentially, whilst in parallel processing each pixel is assigned an individual processor, and all processes operate in parallel. This leads to substantial improvements in computational speed (Fountain, 1983).

An alternative to template matching is to extract measurements or features from the image and to use these in its classification. This is known as *pattern classification* in feature space, a particularly useful technique when there is a variety of objects randomly positioned in the field of vision. Many features of the image can be used for this purpose—area, perimeter, minimum enclosing rectangle, centroid, minimum radius vector, maximum radius vector, holes, etc. A feature vector can be constructed for each object, and this can then be compared with stored feature vectors for the purposes of identification.

The 'bin-picking' problem is one of the major challenges faced by object recognition systems. For a robot to be able to retrieve a particular component out of a bin of mixed components, it must be able to recognize the different pieces, and to determine their relationship to each other and their positions in three-dimensional space. One commercial system thresholds the image into light blobs and homes in on the largest blob, on the assumption that it will be the easiest to grasp. Identification follows later (Edson, 1984). Another system attempts to identify the parts before acquisition by comparing the feature vectors of the light blobs with stored templates.

Depth measurement and analysis

The above techniques were in the main concerned with the recognition of an object from a two-dimensional image or silhouette. The third dimension, that of depth or range, has not been considered. Earlier it was demonstrated how range information could be obtained by optical means using time of flight and optical triangulation methods. Here we illustrate how robot vision systems can determine range and how that information is used in object recognition.

First we consider stereoscopy, or passive triangulation, i.e. triangulation carried out in environmental or ambient light. This technique requires two cameras, whose optic axes converge on a point P which forms images at the

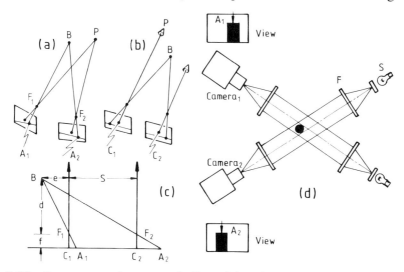

Figure 8.13 Stereoscopy for range finding: (a) point B from images at A_1 and A_2 (b) if the optic axes are parallel the epipolar lines are parallel to F_1F_2 (c) geometry for determining range (d) use of parallel projection optics.

principal points C_1 and C_2 (Figure 8.13(a)). An object point B, not in the plane PC_1C_2, forms images at A_1 and A_2. Camera lenses are at F_1 and F_2. The plane BF_1F_2 intersects the images along epipolar lines, and as point B moves in this plane its projections A_1 and A_2 move along these epipolar lines (Fryer, 1984). It can be seen from Figure 8.13(a) that the epipolar lines are not parallel, so that the stereoscopic disparity of the two images of B has vertical and horizontal components. This complicates the determination of the absolute position of B. Calculations can be considerably simplified if the optic axes are parallel, as in Figure 8.13(b). In this case the epipolar lines are parallel to F_1F_2, and this can be of particular advantage in rectangularly sampled images.

Figure 8.13(c) shows a projection on to the plane of the optic axes. By similar triangles

$$\frac{d}{f} = \frac{e}{C_1A_1} = \frac{e+s}{C_2A_2} \tag{8.14}$$

or

$$d = \frac{sf}{C_2A_2 - C_1A_1} = \frac{sf}{\beta} \tag{8.15}$$

where β is the stereoscopic disparity. It is easy to show that the height H of point B above the plane PC_1C_2 is given by

$$H = dh/f$$

where h is the distance of the image points A_1 and A_2 below the line C_1C_2.

A major difficulty, known as the *correspondence problem*, lies in the identification of the points A_1 and A_2 that correspond to images of the same object point B. This is a problem because certain surfaces, visible from one camera, could be occluded to the other camera. Furthermore, the particular lighting effects may show different image characteristics in the two views; a shadow present in one view may not exist in the other. Correspondence is established by matching specific features such as corners or other clearly defined regions. Range resolution of 5 mm in 2 m has been reported (Yakimovsky and Cunningham, 1978).

Backlighted parallel lighting has been used to circumvent some of these problems (Jones and Saraga, 1981). The task, that of placing three different sized rings on to a tower consisting of three eccentric discs, was intended to contain the key elements of a number of industrial assembly problems. Parallel projection optics were used (Figure 8.13(d)). The object was illuminated from behind by parallel beams generated by small light sources S at the foci of Fresnel lenses F. Combinations of further Fresnel lenses and camera lenses produced real images of the parallel projections of the object on the two television cameras. The image size is independent of the position of the object, and this simplifies the picture analysis. In addition the position of the object perpendicular to the axis is given directly by the position of its image on the television camera. A further advantage is the immunity to changes in ambient illumination, since only light which is clearly parallel to the optic axis passes through the camera lens.

Active, as opposed to passive, triangulation uses an actively controlled light source to illuminate selected points in the scene, and a directional sensor that can detect the reflectance of the controlled light source. In general, no special ambient lighting conditions are required. Two optical range finders of this type were referred to earlier (Figure 8.10). The principle is essentially the same as stereoscopy (Figure 8.13(a)), with one of the cameras replaced by a collimated source of light. A decided advantage, however, is the elimination of the correspondence problem.

The scanning of an object can be speeded up if a plane of light is used instead of a single beam. This form of illumination is known as structured light; the information contained in the structure of the light is combined with that in the image to determine range. The principle is illustrated in Figure 8.14(a). The object is illuminated along a plane in three dimensions, and forms a line image on the imaging plane. Each point along the image line is associated with a straight line to the corresponding object point. The object point is situated at the intersection of this straight line and the known illuminating plane.

This form of structured light has been used in a vision-based robot system designed for picking up parts randomly placed on a moving conveyor belt (Holland *et al.*, 1979). That application, however, was not concerned with range information, i.e. the *z*-dimension in Figure 8.14(a). It used the deflection of the line *CD* to determine the geometry of the component in *XY*-coordinates. A linear array camera was positioned vertically above line *AB*, the projection of the light plane on a horizontal moving belt. When an object intersected the beam the projection was displaced to *CD*, out of the

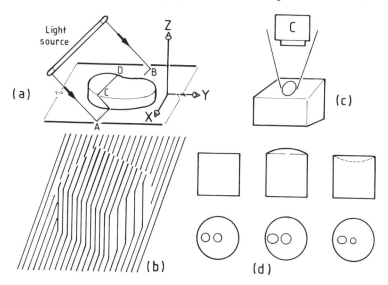

Figure 8.14 Structured light for range finding: (a) deflection of line *CD* by object is used to determine its shape (b) use of parallel planes of light (c) use of a cone of light (d) camera view of objects with similar plan view.

camera's field of view. Thus the camera saw a bright line with a gap in the middle equal to the width of the component. The camera scanned the belt at a constant rate, and for each equal increment of belt travel the vision system recorded one of these scans. This provided sufficient information to determine the object's geometry in XY-coordinates. Planar structured light is also used to guide welding robots. The intersection of the sheet of light with the workpiece surface forms a curved or straight-line stripe that depicts the layout of a particular cross-section of the workpiece. This information can be used to guide the welding torch (Morgan *et al.*, 1983).

A plane of light may be obtained by illuminating a linear slit or by passing a collimated beam through a cylindrical lens (Nevatia, 1982). Figure 8.14(b) shows the image formed by scanning a cube with a series of parallel planes. As can be seen such a procedure gives poor resolution for surfaces nearly parallel to the illuminating plane, but this can be counteracted by using two mutually orthogonal series of scans (Shirai, 1972), a technique easily implemented by using a scanning mirror and a rotatable cylindrical lens. A further extension is to use grid coded structural light, in which a square grid is projected on to the object's surface (Nevatia, 1982).

A final example of structured light is illustrated in Figure 8.14(c). In this case the light is structured in the form of a cone which, when projected downwards on to the object's surface, forms a bright line enclosing an area whose shape is determined by the nature of the reflecting surface. For example, if the surface is flat and at right angles to the axis of the cone, a circle results; if it is flat and angled an ellipse results. These images are received by a camera, collinear with the axis of the cone, this arrangement eliminating the effects of shadows. Figure 8.14(d) shows the images received from three objects which differ only in their top surfaces and which would be difficult to separate if viewed from above in unstructured light.

Variations in reflectivity with distance have also been used for range detection (Nevatia, 1982), but space does not permit further discussion here. It is now time to examine how the robot can use its sensory information in an 'intelligent' way.

8.7 Artificial intelligence

Alan Turing, one of the pioneers of computing, proposed what has become known as Turing's interrogation game as a definition of artificial intelligence. The idea behind this definition is that of the party game in which a player, by having a conversation with a person, has to determine whether that person is a man or a woman. A third person acts as an intermediary so that the player cannot see or hear, but gets all the information by asking questions. Turing suggested that if a similar game were set up with a machine, and the player could not tell whether the hidden 'opponent' were a human being or a machine, then it could be asserted that the machine had achieved the level of intelligence of the human being.

Artificial intelligence has two thrusts: to understand the principles of intelligent behaviour, and to build working models of intelligent behaviour (Winston, 1977). This book is concerned with the latter, particularly in relation to intelligent robots. The field of artificial intelligence comprises a broad range of research activities which overlap to greater or lesser degrees (Barrar, 1980), including visual perception, language understanding, expert systems, data management, automatic programming and game playing. Of these the first two are probably of the greatest relevance to robotics. We shall here look more closely at visual perception. Voice input, a subset of language understanding, will be covered in Section 10.5. (See Winston (1977), Barrar (1980), Winograd (1972), Baker (1981)).

Visual perception

The goal of providing a robot with its own means of perceiving the world around it is an important one—one whose achievement would allow robots to act as inspectors, as assemblers, as explorers. Indeed it is a necessary step along the road to the autonomous robot.

As we have already seen, computer vision can be used for the identification of components; that in itself could be claimed to be an elementary demonstration of machine intelligence. The task of identification is essentially one of mathematical pattern recognition which uses various features computed from the pattern. *Scene analysis,* the next stage in complexity, aims to break down a scene into its constituent components and to determine the spatial relationship of these components.

Because of the complexity of this task, much of the relevant research to date has concentrated on a 'blocks world', consisting of polyhedral objects on a table-top. Three-dimensional scene analysis originated in the early 1960s and was pioneered by Roberts' work (1963) on the recognition of three-dimensional polyhedra from a single view. His programme allowed dimensions, location and orientation of objects to be determined, and armed with this information it could construct an image of the scene observed from another viewpoint. Edges and corners were detected first and used to construct a line drawing or primal image. This image was then interpreted by identifying triangles, quadrilaterals and hexagons, which suggested possible outputs, and eventually accounted for all lines and junctions as edges and corners of particular objects.

A simplified account of topological matching, based on Roberts' method, is described by Aleksander and Burnett (1983). Figure 8.15(a) shows a primal image of a cube. By numbering the vertices 1 to 7 (8 is hidden), it is possible to present the topology (or structure) in matrix form, as in Figure 8.15(b). This shows a 1 wherever two vertices are joined by an edge, and a 0 otherwise. Similar matrices can be constructed for other polyhedra, viewed from different angles, and their use allows comparison with a series of stored matrices, each representing a particular polyhedron. Recognition of a particular block, based solely on its topology, is therefore possible.

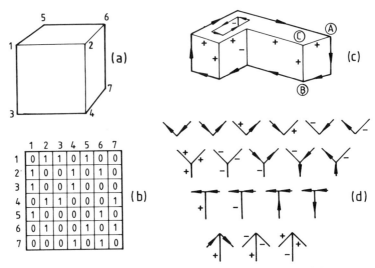

Figure 8.15 Visual perception: (a) image of a cube and (b) its matrix representation (c) edges and lines (d) the 18 physically realizable vertices for trihedral objects.

As the number of objects in a scene grows, the occlusion of some by others becomes a major problem in terms of recognition. This can be resolved however, if the lines, vertices and faces belonging to the different objects can be separated (Nevatia, 1982). This, known as the segmentation problem, has been the subject of much research. (See Winston (1977), Guzman (1968), Huffman (1971), Clowes (1971)).

Interpretation of a line drawing of a three-dimensional object requires a study of lines and how they combine at vertices. Lines fall into two major categories, boundary lines and interior lines, with boundary lines separating regions belonging to different objects, and interior lines separating regions belonging to the same object (Winston, 1977). Interior lines can be further divided into convex edges and concave edges. In convex edges the two faces recede from the edge from the observer's point of view: in concave edges the faces are towards the observer. This is illustrated in Figure 8.15(c) in which a + indicates a convex edge, a − a concave edge, and a boundary arrow is directed so that the visible face is always to the right.

We shall restrict ourselves to a consideration of trihedral objects, i.e. those in which each vertex is formed by the intersection of three faces (as in Figure 8.15(c)). For such objects it is easy to see that only four types of vertex are possible, the L, the fork, the T and the arrow (shown in descending order in Figure 8.15(d)). Since there are four ways to label any given line there are 16 ways to label an L, and 64 ways each for forks, arrows and Ts, giving a grand total of 208 theoretically possible vertices. Fortunately it has been shown (Winston, 1977; Guzman, 1968; Huffman, 1971) that only 18 of these are physically realizable; these are shown in Figure 8.15(d).

How is this information used in scene analysis? Reverting to Figure

8.15(c), an L-shaped object with a hole in it, it was easy to label the edges knowing the physical nature of the object. The trick is to reverse this process so that the physical reality can be inferred from a knowledge of the edges. Let us attempt to do this assuming that Figure 8.15(c) is a primal image. First, the border separating an object or collection of objects from a background is a closed loop with arrows around it in a clockwise direction— eight lines are thus identified. The edges of the internal hole must form a closed loop with the labelling arrows traversing an anticlockwise route—a further four lines are thus identified. Eight vertices have now to be examined. There are three forks and five arrows. Considering the arrow at vertex *A* we note that the bottom row of Figure 8.15(d) indicates that the internal edge at *A must* be convex. A similar argument shows that the internal edge at *B* is also convex. Moving now to the fork at *C* we note that the edges from *A* and *B* are both convex, and Figure 8.15(d) shows that the third edge is therefore constrained to be convex also. Proceeding in this manner allows all edges to be identified. Hence, starting from a two-dimensional line drawing, it is possible to determine the topology of the real object.

This technique has been developed to account for cracks (i.e. where one body rests on another), and shadows (Waltz, 1975), but the major advance was achieved by Winston (1975) who showed how simple objects, once identified, could be arranged in so-called *semantic nets* to represent more complicated objects. For example, his program could examine a blocks world and identify particular arrangements as tables or as arches. This was a big step along the road from seeing to perceiving.

8.8 Recapitulation

There is a need for a second generation of robots with the ability to interact with and to relate to their external world. Such machines require sensing capabilities not unlike those of a human, but in many cases much superior. Two of these senses, touch and vision, have been examined in detail.

Tough sensors can be quite simple mechanical feelers or microswitches, but tactile sensors are more complicated: they can give information on force, position, slip and shape. Touch/tactile sensors respond to reaction forces between a gripper and an object, but gross force/torque sensors provide information on the reaction forces/torques between an object and the external environment. They are important in assembly work.

Non-contact sensing has a number of advantages. Proximity detectors and range finders, for greater distances, have been discussed.

Several aspects of machine vision, including image acquisition and processing, have been described. These, and object recognition, require vast amounts of information to be processed. Depth measurement is particularly challenging: it can be accomplished by passive triangulation, as in stereoscopy, or active triangulation, as in structured light.

The provision of machine senses is the first step towards artificial intelligence: this is illustrated with reference to visual perception.

Chapter 9

Robot applications

9.1 Introduction

Many texts on industrial robots attempt to predict the future capabilities of these machines. Predictions are of two main contrasting types: those which seek to allay fears of robots taking over employment by emphasizing that the abilities of machines are, and always will be, limited; and, at the other extreme, there is the school of thought that, with advances in control technology and artificial intelligence, there is no limit to the tasks that robots may undertake. Such predictions are often proved wrong, and we will not indulge in them except to point out that it is perhaps wrong to blinker oneself by considering manufacturing industry alone. The technologies which contributed to, or sprang from, industrial robot development are finding applications in areas other than the factory floor, and this book attempts to reflect that diversity. However, having said that, this chapter confines itself to industrial robots in manufacturing processes. Applications are many and varied, so in the interests of space it is necessary to concentrate on some of the more common areas.

9.2 Spot-welding

Most welding processes use an electric current to heat and melt the workpieces to be joined. Resistance spot-welding, commonly called spot-welding, welds material by passing a current through the joint. Mechanical pressure is also required to consolidate the joint. The current is derived from a stepdown transformer, and the arms, electrodes and work complete the secondary circuit.

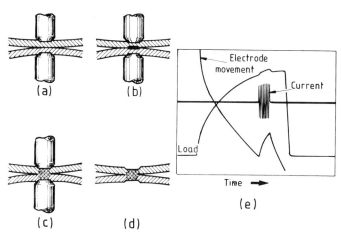

Figure 9.1 Spot welding: (a) pressure application (b) current flow (c) pressure maintained after current ceases (d) finished weld (e) current, load and electrode movement patterns.

The formation of a spot-weld depends on a cycle of events:

(1) pressure is applied
(2) passage of current melts material
(3) pressure maintained while weld solidifies
(4) pressure released

Figures 9.1(a) to (d) illustrate a cross-section of the spot-weld during these stages, and Figure 9.1(e) (Houldcroft, 1979) shows how load, electrode movement and current vary with time. Initially pressure is applied to bring the workpieces into contact. The correct choice of electrode material and form means that the resistance of the dry joint is higher than the other resistances in the circuit—electrodes, electrode/work contacts, work material. The passage of current causes heating at the joint, and a nugget of molten material forms. The nugget is effectively sealed, and its expansion with temperature rise causes a slight opening of the electrodes. After the current has ceased the nugget cools by conduction to the surrounding material, solidifying and contracting; pressure is released and the weld is complete.

A number of parameters affect the size, form and quality of a spot-weld—electrode diameter and shape, force applied, current, timing of events. It is not practicable to monitor size or temperature of the nugget during welding, or to test the weld subsequently. Optimum welding parameters are normally established in advance for particular work, and the only direct control of welding necessary, in manual or automatic systems, is positioning of the electrodes and initiation of the welding cycle. The cycle in Figure 9.1(e), for example, welds two 1.6 mm mild steel sheets by raising the electrode force to 4400 N, passing 5.8 kA for 15 cycles (50 Hz supply) and maintaining electrode force for a further 0.45 s. The electrodes are circular in section with conical or dome shaped ends to ensure good contact, and are usually water-cooled. Clamping force is normally provided by a pneumatic cylinder.

More industrial robots are used for spot-welding than any other application. The automotive industry, always in the forefront of automated manufacturing, was quick to recognize spot-welding as a promising area of application for robots. Spot-welding may be automated with dedicated equipment, and robots share with these systems the advantage of consistent quality of weld, but the inherent flexibility of robot systems is particularly valuable in automobile manufacture, where it is often necessary to accommodate frequent design changes with the minimum of resetting or reprogramming.

There are four factors contributing to the time taken to spot-weld an automobile body:

- The time for each weld, which is fixed and not under control of the robot.
- The time taken in moving the electrodes from one spot to the next. This time is significant since the short distance between spots does not allow the robot to attain maximum speed. The time taken to travel between spots is thus heavily dependent upon the robot's acceleration and deceleration capabilities.
- Motion of the electrodes between welding areas, which may require major reorientations.
- Robot idle time while work is loaded or unloaded.

The production rates encountered in automobile manufacture allow large multirobot installations, so that motion between welding areas can be reduced by allocating a separate robot to each area of body. There is a practical limit, however, to the number of robots which can operate simultaneously on one body; hence the familiar spot-welding line, a typical example being shown in Figure 9.2 (Stauffer, 1983). The line comprises a number of stations, and all robots weld simultaneously. When all the robots have completed their welding tasks, each body is moved to the next station, i.e. the line is indexed, one completed body coming off the line and a new part being loaded onto the line.

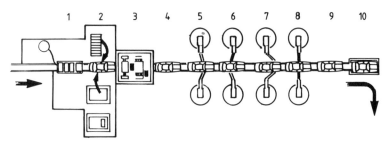

Figure 9.2 Automobile spot-welding line.

The line shown in Figure 9.2 is half of the body framing line of the Jeep Sportwagon made in Toledo, Ohio. This part of the line puts more than 300 welds along the roofline and upper frame (there are approximately 3200 welds in a completed body), and there are ten stations in the line.

The partially built body arrives at station 1, where sealant is applied manually. After indexing, the body is at station 2 where the appropriate roof panel and cowl top are placed on manually, and secured with toytabs. At station 3 a double-gantry robot and fixed welding guns place the initial welds, establishing part geometry. An idle station (4) is followed by four welding stations of two robots each, at which the welding of the roofline and upper frame is completed. After another idle station the assembly is automatically unloaded at station 10. Cybotech model H80 robots are used: hydraulically powered machines with six degrees of freedom. Since electrodes are circular they can be revolved about their axis without affecting the weld, so in some circumstances it is possible to spot-weld with a manipulator having five degrees of freedom. However, six degrees of freedom are more practical in automobile assembly since the angle of approach to some welds is often restricted by the need to avoid collision points between the assembly and the robot arm.

The ability of a robot to generate output signals and to operate conditionally on the status of input signals is essential to the operation of such a line. The last steps in the sequence of each robot are to withdraw clear of the work and to generate an output signal to signify completion of the task. These signals are accepted by a sequencer, which will index the line only when all robots have signalled completion. Each robot is then waiting for a signal from the sequencer to indicate that the line has indexed and welding may commence. The sequencer may be a PLC or computer; in the example quoted above the sequencer is a PLC, and it communicates with the robots via a PLC at each station.

The cycle time of a line consists of the cycle time of the slowest station plus the time taken to index the line. Good balancing of the line ensures minimum robot idle time.

Since the secondary circuit from the welding transformer carries a high current at low voltage, losses can be high in the cable from the transformer to the welding gun if the transformer is mounted beside the manipulator and the cables taken along the arms. This may be overcome by mounting the transformer with the welding gun (Figure 9.3(a)); the flexible electrical connection is lighter and does not sustain such losses since it carries a relatively low current at high voltage. A water supply is needed to cool the electrodes, and air connections are required for the mechanical action. If weight prohibits this arrangement (the mass of a welding gun can be as high a 60 kg) the transformer may be mounted overhead and the secondary current taken through a flexible overhead connection—a 'kickless cable'—to the gun (Figure 9.3(b)).

Bearing in mind the mechanical and electrical losses in the above systems, and that the flexible cables, which restrict the robot's movement,

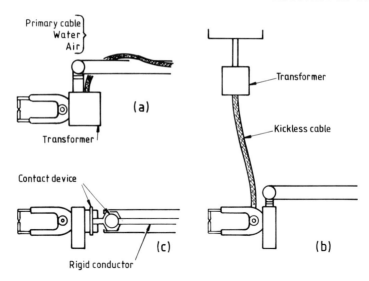

Figure 9.3 Robot spot-welding: (a) wrist-mounted transformer (b) remote transformer (c) secondary current through rigid conductors and contact devices.

have limited life, a system of rigid secondary conductors mounted on the arms has been developed (Figure 9.3(c)) (Kaufmann, 1983). Efficient rigid conductors are used, and joints are accommodated by a contact device which consists of two sheets in sliding contact. The device requires pressure to push the sheets together for a good electrical connection, and this is supplied by a pneumatic cylinder. It is of no consequence that the contact device is mechanically locked when contact is established, since the robot is always stationary when current is required.

9.3 Injection-moulding

Injection-moulding is the process of heating polymer granules to the correct viscosity and metering a 'shot' (i.e. the correct amount to fill the mould cavity) into a mould. The moulding process itself needs only a brief mention since the role of the robot in injection-moulding is that of a machine unloader: its presence does not directly concern the process.

The moulding cycle has six steps:

(1) closing of mould
(2) injection of polymer
(3) dwell time
(4) setting (a heated mould aids curing of thermosetting polymer; a cooled mould stiffens a thermoplastic material)
(5) opening of mould
(6) removal of part

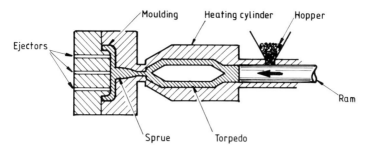

Figure 9.4   Injection moulding system.

Very large forces are required to lock the mould during injection, since the high injection pressure (up to $120 \, \text{MN/m}^2$) may act on a large projected area of the part. The machines therefore, tend to be large and expensive, and it is desirable to minimize machine idle time, i.e. to speed up unloading. When the mould opens, the sprue through whch the polymer entered the mould presents a horizontal tapered bar convenient for gripping the part for withdrawal along the machine axis (Figure 9.4). Three degrees of freedom are sufficient for automatic unloading, and the injection moulding machines are often fitted directly with unloaders which are in effect pick-and-place devices with three degrees of freedom.

It would be uneconomical to employ a flexible robot system in the inflexible role of unloader for a single machine. It has been found feasible to unload up to four machines with one robot, and the parts can be taken directly from the machine to other operations, e.g. trim, drill, palletize or further assembly. A typical example is the manufacture of a plastic vacuum-cleaner housing (Warnecke and Schraft, 1982), where the robot unloads the housing from the injection-moulding machine and places it in a trim press, where flashing is removed. The part is then taken to a punching machine, and finally the robot places the housing on a conveyor.

The interlocking between the robot and other machinery is an important part of such an operation. Consider the cycle in more detail. The protective door on the injection-moulding machine opens automatically when the moulding cycle is completed, and a detector (microswitch or magnetic switch) is fitted to indicate when opening occurs. The output from this detector is wired to an input on the robot controller whose program requires that input to switch on before it initiates the unloading operation. When the robot has taken the part clear of the machine the controller starts the moulding cycle. On some installations, before initiating moulding, the robot may check that the part has been properly unloaded by taking it to a suitable detector (this aspect is described in detail in Section 9.10). This is to prevent the mould attempting to close on solid material stuck in the cavity from the previous operation. Similar interlocks are required for interfacing to a trim press or to any other machine on to which the part is subsequently unloaded.

222   *Robotics: an introduction*

9.4 Arc-welding

In arc-welding, the work, and an electrode held some distance from the
work, form part of an electrical circuit which is completed when an arc
bridges the gap. Heat for melting the metal is generated by the arc.
Arc-welding processes are classified as consumable-electrode and non-
consumable-electrode, and the two processes most commonly used in robot
welding situations—inert gas–metal–arc (MIG) and inert gas–tungsten–arc
(TIG)—fall respectively into these classifications. As the names imply, both
processes are gas-shielded.

The elements of a MIG welding system are shown in Figure 9.5(a). The
electrode is positive, and magnetodynamic forces carry drops of molten
metal from it across the arc to fill the weld pool. The filler wire which
constitutes the electrode is consumed, and must be replenished automatically. This process is particularly suited to automatic operation since a stable
arc is maintained if the wire is fed at a constant rate. The amount of wire
melted in a given time (the *burn-off rate*), depends on arc current, which in
turn depends on the voltage across the arc. If the burn-off rate is less than
the wire feed rate the end of the electrode effectively approaches the work.
This decreasing arc-gap causes a drop in voltage and therefore an increase in
current through the circuit. The burn-off rate thus increases until the arc-gap
is restored to its equilibrium dimension. The essential features of a stable arc
are therefore a constant wire feed rate and a power source with a negative
characteristic.

Wire is taken from a reel which is friction-loaded to prevent snagging
during unwinding. The wire enters the gun end from a copper tube, contact
with which completes the electrical circuit. The gas is introduced around the
electrode, and for currents above about 250 A the gas nozzle is water-
cooled. Currents up to about 700 A are met in MIG welding.

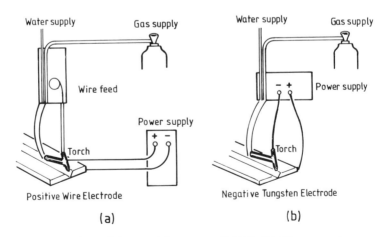

Figure 9.5 Arc welding: (a) Metal inert gas (MIG) (b) Tungsten inert gas (TIG).

TIG welding equipment is shown in Figure 9.5(b). For currents less than 100 A a positive electrode may be used (a positive electrode is desirable since there is an inherent cleaning action on the work). A negative tungsten electrode is usual, however. The arc-voltage/arc-gap relationship is affected by the shape of the electrode—commonly in the form of a cone, the angle of the cone being dictated by the current range to be employed. If the cleaning action of a positive electrode is required at currents over 100 A (which would melt and therefore consume the electrode), AC power sources may be used. TIG welding is a versatile process: currents from 0.5 A to 25 A are used with air-cooled torches to weld thin metal from a negative electrode; from 25 A to 350 A AC power is suitable, and up to 800 A may be used when a negative electrode and a water-cooled torch is used to weld thick metal. If filler is required in TIG welding it is supplied from filler wire introduced directly into the weld pool.

TIG welding is shielded by a flow of inert gas, or nitrogen, or a mixture of these. Argon and helium are common choices, and a purity of at least 99.5% is demanded. As well as the gases mentioned, carbon dioxide is also met in MIG welding.

Figure 9.6(a) shows the main features of a robot MIG welding station for welding steel components. Six elements of a robot arc-welding system have been identified (Nally, 1983):

- Robot (manipulator, power unit and controller)
- Welding equipment (power source, wire feeder, torch interface, cables)
- Work-positioning device (one or more tables)
- Work-holding device (jigs and fixtures)
- Systems engineering (welding program and procedures)
- Filler metal

Each of these elements warrants further consideration.

As with spot-welding, the general observation can be made that in theory five degrees of freedom are sufficient for arc-welding (the electrode could be rotated about its axis without affecting the weld), but six are more practical. Electric and hydraulic manipulators are both used for arc-welding. Since high speeds are not required, welding versions of standard manipulators are often produced in which speed is sacrificed for improvements in accuracy and repeatability. Teaching may be by the lead-through method, where the torch is taken manually through the welding path, and the path is then retraced by the robot. The majority of welds are along straight or circular seams, and a point-to-point control with interpolation (see Chapter 7) is useful, since only the endpoints of each seam need be taught for a straight line, or three points in the case of a circle. In its simplest form, communication between the robot controller and the welding equipment consists of a digital signal sent to the welding controller to indicate start and finish of welding. The parameters of the welding process—wire feed rate, current, gas flow rate, timing of gas on/off relative to current on/off—will

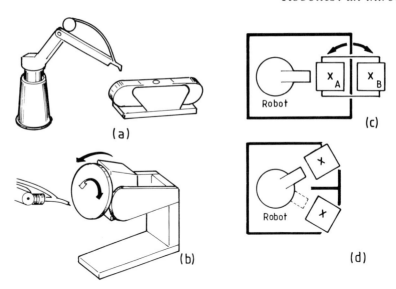

Figure 9.6 Robot arc-welding: (a) with turntable (b) with two-degrees-of-freedom positioner (c) manually loaded turntable (d) two-table arrangement for manual loading.

have been determined and set in advance, and will be controlled by the welding equipment.

As far as the welding equipment is concerned, the power source and gas storage are floor-mounted, and flexible cables and tubes mounted on the arm of the manipulator, or taken from above, supply the torch with current, gas and cooling water. The filler wire must be flexible to allow it to feed from a reel, and it is desirable, to avoid snagging, to keep the distance from reel to torch short. It is usually found practical to mount the wire feeder near the elbow of the manipulator. The torch may be attached to the wrist by a wooden, plastic or metal link designed to shear under excessive strain. This is a precaution against the possibility of damaging the torch if a fault should cause the torch to be welded to the work.

Work positioning devices perform two functions:

- By rotating the work round one or more axes, extra degrees of freedom can effectively be added to the manipulator/positioner combination.
- Multi-table devices exchange work between robot and human operator.

Figure 9.6(b) shows a two-axis positioner; when this is allied to a system with five degrees of freedom, two further degrees of freedom are obtained. The positioner is controlled by the robot computer, and is not usually in motion during welding; rather it is taken to discrete positions for each seam. This means that, for example, internal welds round the base of a rectangular box are easily accomplished. This would not be possible from a single position, even for a manipulator with six degrees of freedom.

In manufacturing operations maximum returns usually require maximum usage of capital equipment: this also applies to welding. It is claimed that robot systems give high welding times compared to manual or semi-automatic systems, but this depends on the minimization of the robot idle time which occurs during loading and unloading.

Figures 9.6(c) and (d) show two methods of co-ordinating loading and unloading by an operator with robot welding. With the double table arrangement in Figure 9.6(c), the robot welds a seam on a part on table *A*, while the operator loads table *B*. When welding is complete the robot computer waits for a signal from a pushbutton or footswitch pressed by the operator, to indicate that loading is complete. The table, which is directly under the control of the robot, is turned through 180°, and when sensors indicate that this rotation has been completed welding is commenced again. Meanwhile the operator is free to unload the finished part and reload with fresh work. There may be variations of this arrangement: four-position tables are met, and if the ratio of loading time to welding time permits, one operator can feed more than one robot. It is particularly easy to screen off the robot/turntable set-up (a requirement for any arc-welding operation), by enclosing the whole cell with a fixed screen. The two halves of the turntable are divided by a screen fixed to the centre of the turntable, which is therefore closed when welding is in progress.

It was stated that current, gas flow, and wire feed speed are controlled by the welding equipment, and are normally determined and set in advance. This is satisfactory if the same parameters are acceptable for all welds on one piece of work, but lacks versatility should two different wire feed speeds, for example, be required on one job. Systems have been developed to give the robot more direct control of the welding parameters, a typical example being that for Puma robots, called the Puma Arc Welding System—PAWS (Pavone, 1983). The elements for the system are shown in Figure 9.7(a). VAL is the Puma's control language, and DIGIMIG is a proprietary welding system. The welding controller has been modified to interface with VAL via an RS422 link (a digital serial communications link).

VAL uses English-like commands to program the robot, e.g. to move in a straight line to a taught point identified as *W*1, the appropriate instruction is MOVES *W*1. Extra instructions for the welding situation have been added to VAL. DIGIMIG includes closed-loop control systems for maintaining desired wire feed speed and arc-voltage, and the set-points to these control systems are received from the robot control. Any errors, for example loss of current, are communicated back to the robot control, where they are interpreted and displayed to the operator. At the robot controller, different sets of welding parameters may be defined, and then called up when needed. For example

$$\text{WELDSET } 1 = 425, 29.4$$

defines a wire feed speed of 425 inches/min (a commonly used unit in welding) and an arc voltage of 29 4. If the torch is at position *W*1, then instructions

$$\text{WELDSET } 1, \quad 18$$

$$\text{MOVES } W2$$

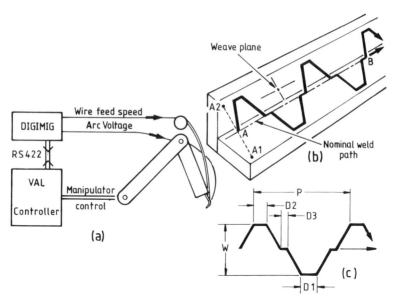

Figure 9.7   Robot arc-welding system: (a) PAWS system (b) actual weld path with
weave (c) specifiable dimensions in weave pattern.

perform a weld with the parameters previously defined in a straight line from
$W1$ to $W2$ at a speed of 18 inches/min. At the end of a weld a crater would
be left if the wire feed and arc stopped immediately the travel had stopped;
this crater is filled by continuing to weld for a short time while the torch is
stationary. For example,

CRATERFILL 0.75, 2

WELDEND

finishes off a weld by filling the crater for 0.75 s at the parameters specified
by WELDSET 2 (which are not necessarily the same as those for the weld
itself).

   For nominally straight welds, better results are often achieved by
weaving the torch along the seam. Figure 9.7(b) shows how the weave
pattern relates to nominal path for a filler weld. The PAWS system allows
definition of the pitch P, dwells $D1$, $D2$ and $D3$, width W and the plane
$A1$–A–$A2$–B by programming and teaching (Figure 9.7(c)).

   Non-sensor-based robots require a higher-quality weld preparation than
manual systems. They cannot, for example, compensate for deviation in
width or position of the seam from the programmed values. Chapter 8
indicated that there is a growing number of second-generation or sensor-
based robots with the ability to deal with such deviations. Various sensors
are employed, including tactile, optical and electrical forms. Two systems
are considered here. An optical system has been described (Kremers *et al.*,
1983) which uses structured light, in this case a single sheet of light, to
illuminate the seam ahead of the welding torch. The view is gathered by a

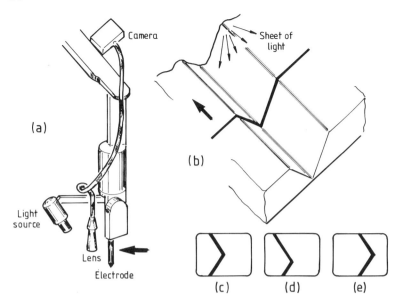

Figure 9.8 Vision-guided robot arc-welding: (a) mounting of light source and camera (b) intersection of plane of light with work (c) seam on nominal line (d) seam to left of nominal line (e) seam away from nominal line.

lens, as shown in Figure 9.8(a), and taken through an optical fibre bundle to a camera. Processing of the scene by a computer indicates deviation of the seam from that desired, and this information is transmitted to the robot so that corrective action may be taken.

Figure 9.8(b) shows how the seam is illuminated by the sheet of light, and Figures 9.8(c), (d) and (e) show the view of a seam to be fillet-welded, with the seam respectively on, to the left of, and further away than the nominal line. It can be seen that the position of the line on the screen, which represents a plan view of the line of intersection between the sheet of light and the surfaces of the parts to be welded, indicates the type and magnitude of the deviation. Problems in these types of systems include: smoke obscuring illumination from the source and the view of the cameras, and spurious light from spatter and reflections.

These are overcome by using a powerful source of illumination and special image-processing algorithms to isolate the line of interest from other light spots in the picture.

In the arc-welding process, there is a known relationship between arc-voltage and arc-gap. This fact has been used to develop a sensory system which deduces the seam geometry from the variation in arc-voltage as the torch weaves (Cook, 1983). If the seam is exactly on the nominal line, the arc-voltage varies evenly as the torch weaves; a deviation is reflected in an imbalance of voltage variations which may be interpreted, knowing the relationship between arc-voltage and arc-gap, as a distance correction to be transmitted to the robot control.

9.5 Surface coating

This category of application includes such operations as painting, glazing, applying car-body underseal and metal plasma spraying. Surface coating normally employs lead-through programming since the operation is easily reproduced manually, but is not easily defined numerically.

A typical application is the glazing of washbasins (Warnecke and Schraft, 1982), shown in Figure 9.9. A conveyor line, loaded manually, is indexed round under control of the robot computer. At the painting position the robot has control of a 180° turntable to gain access to both sides of each workpiece. The painting area is protected by a spray booth which serves two purposes: operators are shielded from enamel spray, and material sprayed past the parts is collected and recycled. In this application the robot is an hydraulically powered machine built specifically for surface-coating applications. This is common in the painting field, where the load-carrying capacity and accuracy of a general-purpose robot are not really necessary. In addition to lead-through teaching, painting robots must, however, have large reach and the ability to synchronize movements with a moving conveyor. The painting of automobile bodies illustrates the need for synchronization (Figure 9.10) (Akeel, 1982). Car bodies of different design arrive randomly, hanging from a continuously moving overhead conveyor. A machine vision system first of all identifies each part and communicates the information to a central supervisory computer, which downloads the appropriate program to each robot. Because each individual painting program is taught using a stationary model, the line must be tracked, since it does not move at a precise speed, and the program modified to follow the part. (This aspect was covered in detail in Section 7.2.) Note that the whole painting operation is contained in one large spray booth, thus protecting operators from spray and facilitating recycling of painting material. The painting material may be

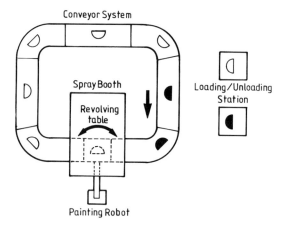

Figure 9.9 Manually-loaded robot enamel-coating system.

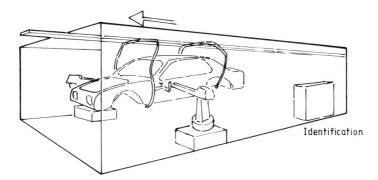

Identification

Figure 9.10 Spray booth for robot painting of automobiles on continuously moving overhead conveyor, with part identification.

liquid or powder spray; electrostatically charging the spray encourages it to stick to the earthed part, which restricts overspray to around 5%. In this particular installation, programs are taught using an identical robot installed elsewhere, working on stationary parts. This permits verification of programs without interrupting production equipment. Programs may also be taught by means of a teaching arm, but this requires the use of a production machine to verify the program.

9.6 Machine-tool servicing

The efficient use of capital equipment is a recurring theme in the discussion of robotic applications, and the area of machine-tool servicing is no exception. The efficient use of a machine tool requires it to be loaded and unloaded as quickly as possible so that machining can occur with the minimum of idle time. The use of a robot for loading and unloading, whilst utilizing the machine-tool to the full, might however incur a long robot idle time. This dilemma is resolved by placing the robot in a work cell. The example shown in Figure 9.11(a) involves two NC grinders (Gandy, 1983), and incoming and outgoing conveyors for new and completed parts respectively. The loading and unloading of both machines keeps the robot fully occupied.

Much of the inherent inspection involved in human operator/NC machine interaction must be formalized for robot operation. Consider, for example, the servicing of an NC lathe. An operator would know from a lightbulb indicator that turning was complete and would then open the protective door, grasp the part and release the chuck. A new part would be loaded, the chuck closed, the door closed, and turning initiated by a pushbutton. Certain inspection and remedial actions are implicit in the operation; for example, swarf clogging a chuck would be brushed aside before loading a new part.

Robot servicing of the same machine must be defined more precisely. In

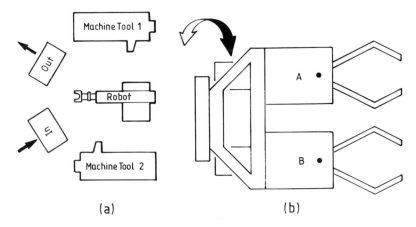

Figure 9.11   Robot work cell: (a) machine layout (b) double gripper.

its simplest form, communication between each NC machine and the robot can be by digital inputs and outputs. The cycle of operations between robot and NC lathe might then be:

(1) signal from lathe to robot to indicate completion of turning
(2) signal from robot to lathe to open door
(3) robot awaits indication from sensor that door has successfully opened
(4) robot grasps part, monitoring gripper sensors to indicate successful grip
(5) signal from robot to lathe to release chuck
(6) robot awaits indication from sensor that chuck has successfully opened
(7) part is withdrawn and taken to next operation (another machine or outgoing conveyor)
(8) signal from robot to lathe to activate an air spray to clear swarf
(9) a new part is collected, gripper sensors again being monitored, and taken to chuck
(10) signal from robot to lathe to close chuck
(11) robot awaits signal to indicate successful closure of chuck
(12) gripper is released and withdrawn
(13) signal from robot to lathe to close door
(14) robot awaits indication that door has closed
(15) signal from robot to lathe to initiate turning

Even a simple loading/unloading operation involves six digital signals from robot to lathe and six signals in the opposite direction (including gripper sensors).

A key feature of many machine-tool servicing applications is the double gripper. Figure 9.11(b) shows such a device for cylindrical parts; it consists of two identical grippers disposed at 180° to each other, mounted directly to the wrist-plate of a robot with six degrees of freedom, or to a five-degrees-of-freedom robot via a 180° turning mechanism. Consider a new part held in gripper *A*. The robot is able to unload the lathe with gripper *B*. The wrist is then revolved 180° in order to present the new part in gripper *A* to the chuck. Unloading and loading are thus achieved in one visit, thereby minimizing machine idle time.

Three-jaw chucks are self-centralizing, so it may not be practical to mount the gripper directly to the wrist-plate: a slight misalignment of the part in the gripper would result in the chuck trying to force the robot out of position. For this reason grippers are sometimes mounted on spring or rubber buffers. Powerful robots are needed for double-gripper operations, since the manipulator must bear the weight of the whole gripper mechanism and two parts.

9.7 Assembly

Although not the most common robotic application, assembly, being very labour-intensive, is the application with the most potential. Manual assembly is often classed as an 'unskilled' operation, but as far as a robot is concerned it is extremely demanding: it requires sensory feedback. First-generation robots did not make much impact on assembly automation, but later machines, with their greater accuracy and better provision for sensory data processing, are remedying this situation.

Chapter 2 described the remote centre compliance (RCC) and its use in the 'peg-in-the-hole' type of task. This principle has been extended in the Selective Compliance Assembly Robot Arm (SCARA) (Makino and Furuya, 1982), a configuration particularly suitable for the manipulator of an assembly robot. In its original form the SCARA had four motors: position of the shoulder and elbow joints defined the coordinates of the wrist in the horizontal plane, and two motors at the wrist controlled the orientation and height of the end-effector. With four degrees of freedom, the machine was restricted to table-top assembly with vertical and horizontal operations. It has been estimated that this covers 80% of all assembly work. A unique feature of the SCARA configuration is that compliance can be controlled—*selective compliance*. Since the arms of the manipulator are relatively stiff, the compliance arising in the joints can be controlled by altering the parameters of the machine coordinate control systems. The SCARA configuration allows continuous rotation of the wrist, making it suitable for drilling and screw insertion tasks without the need for additional motors. Indeed, one of the first uses for the Picmat SCARA was the insertion and tightening of the many screws in a door sash assembly (Figure 9.12(a)).

Another configuration of robot which has proved useful for assembly work is the overhead gantry arrangement. An example (Figure 9.12(b)) is the Olivetti SERIE 3 robot, which can be a multi-arm machine. Machines may be grouped over a conveyor, with work usually mounted on standard pallets which are clamped for accurate location at each assembly station.

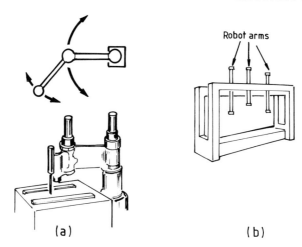

Figure 9.12   Assembly robot configurations: (a) SCARA manipulator (b) multiple overhead arm system.

The applications quoted for this robot (Ruder, 1982), for example electronic connectors, engine covers, injection valves, typify many assembly tasks where the parts to be assembled consist of large parts or subassemblies arriving on a pallet from previous operations, and small fixing pieces, e.g. screws or washers. It is usually found practical to supply large parts and subassemblies to a station on pallets. Small parts like screws are better handled by bowl or magazine feeders. Sensing is an essential feature of successful assembly, e.g. in the single operation of inserting a screw the robot needs to be able to sense the successful gripping of the screw, the jamming of an air-driven screwdriver, the presence of jammed or torn male or female threads, etc. These sensing operations are taken for granted in a manual system, but in an automatic system they require sensors and appropriate hardware and program instructions.

9.8   Packaging and palletizing

A survey of current applications in the packaging and palletizing field reveals a very varied choice of materials or parts to be handled, and many different ways in which parts are stacked or presented to the robot. The complexity of the robots used ranges from pick-and-place devices with two degrees of freedom to controlled-path robots with six degrees of freedom. A prominent feature of all packaging and palletizing applications is the purpose-designed gripper, and two categories of application arise: either parts are presented to the robot in a known, orderly fashion, or the method of presentation is, in some way, random.

Purpose-designed grippers are often required for these applications. Consider the example of a large vacuum gripper. Paper bags containing plastic granules arrive at random intervals on a conveyor, and the robot is required to stack them on a pallet. The bags are hard to handle mechanically since they are not rigid, and a tight grip damages the bag. When air is evacuated from a flexible container of loose particles the packaging becomes quite rigid, and this phenomenon is used to effect here. When a bag arrives at the end of the conveyor it contacts a switch instructing the robot to bring the vacuum gripper to bear on the bag. Vacuum is then applied for long enough to make the bag rigid, allowing it to be grasped and lifted.

A packaging or palletizing application may not always be justified for robotization on its own, but where a robot is introduced into some part of the manufacturing process, it is relatively inexpensive to extend its duties to palletize parts at the end of the process. Many palletizing applications are found in this context, and an example has been cited by Campbell *et al.*, 1984. Here transformer parts are taken from a conveyor to a coating machine, where the part is rotated by the robot under an insulating epoxy dispenser. The robot places the part in an oven, where the epoxy is cured, and finished parts are collected from the oven and stacked in an orderly fashion.

The ability to relocate sequences speeds up the teaching of palletizing operations. If parts are to be stacked in identical layers it is not necessary to teach each part position in the whole stack to the robot separately; rather it is sufficient to teach one layer, and program the heights of all layers—the sequence for loading one layer can then be relocated to the next position up as each layer is completed. Variables are set during the program as a memory of the current stacking position, so that the robot may leave the stacking sequence at any time to attend to other machines in the cell, and return to continue stacking.

A pallet shuttle ensures continuous operation of a palletizing application. One pallet is loaded while an empty pallet waits on another track. When the first pallet is full loading can commence on the empty pallet, and provided the full pallet is replaced by another empty one before loading of the second pallet is complete, the operation of the cell is not impaired.

9.9 Press operations

Three major operations may be accomplished by pressing: cutting, forming and assembly. The first two are confined to producing components from sheet metal, while the third covers a variety of components. Pressing requires the application of a compressive force, usually in the vertical direction, by mechanical or hydraulic means. The role of the robot in press operation is basically that of a loader and unloader.

Firstly we summarize the pressing operations where robots are applicable. Blanking and piercing are distinguished by the fact that blanking cuts the profile of a part, while piercing cuts holes in a part. A punch is forced against the metal strip, cutting a blank into a stationary die. The stripper plate separates the strip from the punch on the return stroke. Consider also the forming process. Bending produces one or more single-axis bends along the entire length of the sheet; the tool is usually set to a few degrees more than the desired angle in order to account for slight elastic recovery on release. By clamping the sheet, a press can be used to stretch-form material, and if the shape requires more deformation, the deep-drawing process is used. Various assembly operations are suitable for presses; a typical example is an automotive wheel hub and bolt assembly, where the bolts are a press-fit into the hub.

It is possible to service press-tools using only three degrees of freedom. This requires the robot to be mounted at the appropriate position with respect to the die and the sheet to be gripped with the correct tool orientation (the roll axis being normal to the sheet). Mechanical grippers are encountered, but vacuum or magnetic grippers are more common in press work.

Manually serviced press machines require thorough safety interlocking. The ram must not begin its stroke until both switches of a two-handed safety control are pressed, and the protective guard is fully down. Provided that the whole robot cell is safely guarded some of the precautions local to the press machine can be eliminated, but interlocking is still required to ensure that pressing does not commence until the gripper is clear of the machine, and that the robot will only enter the machine when the ram has fully retracted. Sensors are required on the gripper to check if the die has been fully unloaded of products and scrap material.

Parts produced by pressing often require more than one operation, so a flexible press machine accommodates more than one die for successive operations on each part. The current trend is towards flexible machines which offer a fully automatic system handling multiple press operations, and the ability to change to different batches of products without the need for manual die-changing. One such machine incorporates a robot, accepting stock from an automatic feeder which rises to keep the top of the stack level, and loading and unloading four dies on the press (Mizutame *et al.*, 1984). Dies can be changed automatically from a stock of 16, and conveyors are built in for the automatic collection of products and scrap.

Some special problems arise in servicing bending operations. Bending is performed by a press-brake, and since large sheets of material may be involved it is important to ensure that the robot can load and unload the press-brake without collision.

9.10 Die-casting

Pressure die-casting is a process for producing parts from aluminium, magnesium or zinc alloys, with high dimensional accuracy and good surface finish. Molten metal is forced into a mould under pressure (25–$200\,\mathrm{MN/m^2}$), and allowed to cool and solidify while the pressure is maintained. Consider the basic elements of the die-casting machine. A heated reservoir is the source of molten metal, and a shot is metered into a cavity formed between two dies. After solidifying the movable die retracts to reveal the casting complete with sprue. The casting is removed, sometimes with the aid of ejector pins, and the dies may be sprayed for cleaning and lubrication before the next casting.

Excess material is found on a raw casting, where molten metal has been forced into the gap between the flat mating surfaces of the dies, and this 'flashing' is removed by a trim press. To make the flashing brittle and easier to remove, the casting is quenched in cold liquid immediately after removal from the die-casting machine. The sprue may hold more than one part, in which case the trim press also performs the function of removing the parts from the sprue and connecting runners. The flashing, sprue and runners are recycled to the die-casting machine.

Consider a robot die-casting cell. The sequence of operations in one cycle of the cell is:

(1) a casting is made
(2) dies part
(3) robot grips sprue and removes part from machine
(4) sprayer cleans and lubricates dies while robot quenches part
(5) part is taken to sensing station for inspection; if in order the die-casting machine is signalled to start
(6) part is loaded into trim press
(7) trim press operates and ejects parts while robot returns ready to enter die-casting machine when next casting is made

Similar requirements exist in die-casting and in injection moulding, as far as interlocking with other machinery is concerned. After quenching the parts are presented to a sensing station, which is an array of infrared reflecting light switches: the program will only initiate another casting if all switches detect metal, otherwise an alarm is activated. (The dies may be accessed manually from outside the cell.) A horizontal jointed-arm configuration of robot (similar to SCARA) in an installation allows the machines to be grouped very closely around the robot. The machines themselves and safety barriers form a closed cell.

Some variations on the installation described are possible. Where cycle times allow, it has been found possible to serve two die-casting machines with one robot. Another possibility is that instead of providing a separate automatic spray unit for cleaning the dies, the spray-gun can be lifted and activated by the robot gripper. The waste from the trim press can be recycled to the die-casting machine automatically using a gravity chute or conveyor. Spraying the dies may not be necessary after every casting, in which case an instruction is entered into the robot program to count the number of cycles of operation and signal for a spraying operation at regular intervals.

9.11   Inspection

There are many applications such as assembly, arc-welding and die-casting where inspection is a necessary and integral part of the process, one that must be carried out during the operation cycle. This section is concerned with automatic inspection systems where the essential features are a robot, an automatic sensory system and a means of communication between the two. Robots and sensors come together when it is necessary to transport the sensor to the part, or the part to the sensor. We have seen an application of the latter in the die-casting example above. Like assembly, inspection covers a variety of processes, and is best illustrated by a study of some specific examples.

A robot has been used successfully to carry an ultrasonic probe used in the inspection of carbon fibre composite parts made in the aerospace industry (Campbell *et al.*, 1984). This material is constructed of a number of plies of material woven from carbon fibre bonded together in a resin matrix. When cured there should be no voids or gas bubbles in the matrix, or any separation between plies, known as delaminations. The non-destructive method of testing for these flaws is to transmit an ultrasonic signal through the material, the presence of voids or delaminations being identified from the pattern of the attenuated signal picked up by a receiver on the other side of the material.

Inspection of a composite part requires a regular scan, usually in the form of a grid, and this is easily achieved for relatively flat parts by three-axis machines built specifically for this purpose. Since it is necessary to keep the scanning head normal to the part surface, at least five degrees of freedom are needed for curved parts, so in this system a robot is used to carry the scanning head. A long caliper configuration of the head mounting allows access to deep components. The abilities of first generation robots can often be greatly enhanced by linking them to microcomputers (Harris and Irvine, 1984), and this example illustrates the point. It is not necessary to program the robot path by teaching, since design information about the shape of the part is available from a computer-aided design (CAD) database. The microcomputer performs two functions: it gathers design information from the CAD system and transforms it into coordinate data suitable for transmission to the robot, and also interprets the information from the ultrasonic system for display on a two-dimensional representation of the part. Synchronization of the system is achieved by the robot signalling, at the start of each line scan, to the computer to start sampling the sensor signal. The computer samples at a rate calculated to monitor points at 3 mm intervals.

9.12   Recapitulation

Current industrial robot applications are many and diverse; in this chapter only some of the more popular areas have been detailed. Computer control is an essential part of most processes, since decision-making and communications with other machines are required. Purpose-built robot systems are available for applications like surface coating; for other applications the choice is wider. Many processes require a specially designed end-effector, and the ability to handle sensors is fundamental to the application of robots to more complex tasks.

Chapter 10

Tele-operators

10.1 Introduction

There are many different definitions of the term tele-operator, and many different words have been coined for these devices, such as telechirs, telefactors, telepuppets and so on. We hope we do not add to the confusion by proposing our own definition. It is not entirely original, being a hybrid of the definitions published in the first major book (Johnsen and Corliss, 1969) on the subject and in reports of the U.S. National Aeronautics and Space Administration (NASA) (Onega and Clingman, 1972). Our definition is that a tele-operator is a cybernetic man–machine system designed to augment and extend human senses and dexterity. Figure 10.1 is intended to clarify this definition. The words have been chosen carefully. 'Cybernetic' implies the concept of control and feedback of information. Its use excludes preprogrammed machines such as automatic coffee dispensers, weighing machines and timer-controlled ovens—and even today's industrial robots. The term 'man–machine' emphasizes the fact that the human element is in control. For this to be the case, the operator needs to have information about what is to be controlled, and this is supplied by feedback from sensors in the work space which augment the human senses. Finally, 'dexterity' means skill or expertise in movement and action, and although strictly speaking it refers to the use of the hands, we here extend its use to the feet. Thus tele-operators augment and extend manipulative and 'pedipulative' human skills. This excludes such machines as automobiles and radio-controlled aircraft.

The prefix 'tele' describes the ability of such systems to project human senses and dexterity across barriers resulting from distance, the hostility of

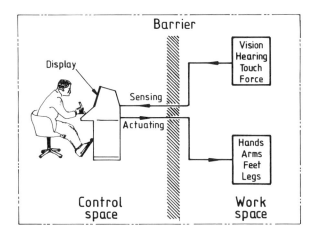

Figure 10.1 A tele-operator augments and extends man's senses and dexterity.

an environment or the physical magnitude of the task to be done. For example, Figure 10.2 shows a selection of tele-operators in which the nature of the barrier differs. In Figure 10.2(a) the control space and the work space could be connected by current-carrying wires, by mechanical cables or by radio waves. Distances could vary from the few metres involved in the manipulation of radioactive materials in a 'hot' laboratory to the millions of kilometres involved in collecting samples of soil from distant planets. In both these cases the hostile environments would present additional barriers.

Distances between control and work spaces shrink to a few centimetres in man-amplifiers and orthotic devices (Figure 10.2(b)) where exoskeletal structures are used to enhance the strength of a normal human or to help a weakened or atrophied part of the body to gain strength and dexterity. In such cases the barrier essentially relates to the physical magnitude of the task. The same can be said of prosthetic devices (Figure 10.2(c)), which attempt to replicate the functions of missing parts of the body.

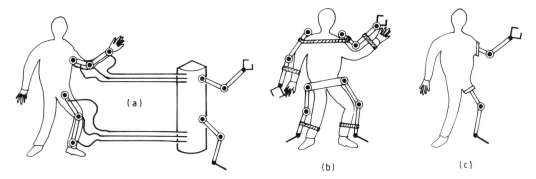

Figure 10.2 A selection of tele-operators: (a) transmission over long distances (b) man-amplifiers (c) prosthetics.

Early applications of tele-operators

Figure 10.3 shows an example of a tele-operator used for handling radioisotopes and radio chemicals, for reprocessing and fabricating nuclear fuels and for inspecting radioactive equipment. Tele-operators of this type were developed about 40 years ago at the Argonne National Laboratory (ANL) of the U.S. Atomic Energy Commission (Goertz, 1952; 1964). As the figure shows, the input, or master arm, is mechanically or electrically coupled to a geometrically identical or similar output or slave arm. In the earliest models this coupling was one-way, from input to output, the slave being made to follow the master. These were known as unilateral tele-operators. Later developments led to the bilateral tele-operator in which the coupling was two-way and the output could drive the input, as well as the other way round. This allowed inertia and work forces to back-drive the master arm, giving the operator a sense of feel of what was happening at the slave end. These bilateral systems were sometimes referred to as force-reflecting teleoperators. Their performance is superior to the unilateral variety.

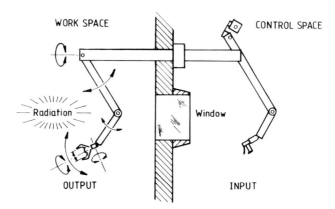

Figure 10.3   A tele-operator for handling radioactive materials.

10.2   Rate versus position control

In extending human dexterity, it is clear that tele-operation requires some means of controlling an end-effector's position and orientation. The human operator plays an important role in this control system, as we can illustrate by considering a one-dimensional task (Figure 10.4). Knowing the target position X_{iw} of a workpiece, it is the operator's job to devise a method of matching the end-effector's position X_m in the work space to this input X_{iw} (Figure 10.4(a)). The operator develops an error-correcting strategy by visually determining the error and mentally assessing various functions such

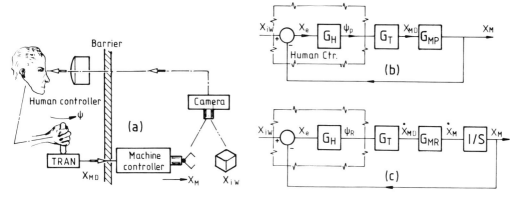

Figure 10.4 Controlling the end-effector: (a) a simple one-degree-of-freedom task (b) block diagram of a position controller (c) block diagram of a rate controller.

as integrals and derivatives. This strategy leads to the generation of a control signal Ψ which demands various positions X_{md}, rates $\dot{X}_{md}$ and accelerations $\ddot{X}_{md}$ of the end-effector. The signal Ψ could be a force for pushing a button in an off–on digital controller, or for pushing against a stiff spring in a proportional mode. It could also take the form of gross displacements such as the rotation of a knob. These transducers (G_t), whether simple buttons or knobs or the more complicated hand controllers (see later), convert Ψ into a form suitable as an input demand to the machine controller (G_m).

Tele-operators use two main types of machine controller: *position controllers* (Figure 10.4(b)) and *rate controllers* (Figure 10.4(c)). When position control is used, the machine controller's output position X_m is proportional to the input X_{md}. When rate control is used the machine controller produces an output rate $\dot{X}_m$ which is proportional to a demanded $\dot{X}_{md}$. Rate control places a greater burden on the human controller; with position control the controller has only to generate a signal Ψ_p which is proportional to the desired position of the end-effector, but with rate control it is necessary to devise a signal Ψ_r whose integral is proportional to the desired position.

Figure 10.4 illustrates a particularly simple tele-operator with one degree of freedom, but it is easy to see how this basic concept can be extended to the six degrees of freedom required for complete control of an end-effector. A separate controller is required for each degree of freedom. However, in extending the degrees of freedom an increasingly difficult task is being presented to the human controller, and it is important that this task be kept within manageable proportions.

An added complication arises when there is no simple relationship between the world coordinates of the end-effector and the machine coordinates determined by the outputs of the machine controllers. This happens, for example, when articulated arms are used. Here the human operator has to match a set of inputs in world coordinates by controlling the machine coordinates and their derivatives. In order to do this the operator must mentally transform the desired position or rate of the end-effector to desired angular positions or angular rates of the joints in the manipulator arm, depending on whether a position-control or a rate-control mode is employed. The combined task of transformation and control of several machine coordinates can be daunting, but it is possible. An excavator arm, for example, is a rate-controlled system with three degrees of freedom, although it does not demand extreme accuracy. In those systems where accuracy is important the operator will often simplify the task by driving one machine coordinate at a time, but this can be time-consuming and it is often necessary to use computers to perform the necessary coordinate transformations.

A tele-operator employing position controllers and a computer for coordinate transformations is shown in block diagram form in Figure 10.5(a), where a bar above a variable indicates a vector. The operator's outputs $\bar{\Psi}_p$ are spatially related to the outputs $\bar{X}_m$, i.e. if a movement is required in the x-direction then $\bar{\Psi}_p$ is moved in that direction. Now since the position controllers G_{mp} act on the machine coordinates $\bar{\Theta}_m$ it is necessary to transform the operator's demanded $\bar{X}_{md}$ into demanded $\bar{\Theta}_{md}$, which are then used as inputs to the position controllers. This is done by computer. The inverse transformation T^{-1} is derived from the Denavit–Hartenberg matrices, as described in Chapters 2 and 4. The final transformation from machine coordinates back to world coordinates is accomplished by the articulated arm itself. The reader will recognize this as a form of geometric control.

The system of Figure 10.5(a) takes the burden of coordinate transformations away from the operator who is nonetheless left with the task of controlling up to six degrees of freedom. This is a manageable task, although in many instances the operator will first position the end-effector using the three translational degrees of freedom before using the rotational degrees of freedom for orientation.

Coordinate transformations can also be handled by the computer when

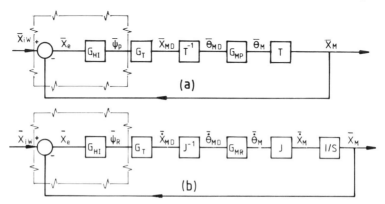

Figure 10.5 Computer-aided tele-operation: (a) block diagram of a system using a position controller (b) block diagram of a system using a rate controller.

rate controls are used (Figure 10.5(b)). In this case the inverse J^{-1} of the Jacobian has to be calculated. This method of control is known as resolved motion rate control (RMRC). The rate controllers drive the end-effector at a rate proportional to, and in the direction of, the signal applied to the input transducer. The RMRC requires less computer effort than the position control system of Figure 10.5(a), since calculation of J^{-1} is easier than that of T^{-1}. In terms of accuracy and time to complete a task, it still has the disadvantage of requiring the operator to generate a signal vector $\bar{\Psi}_r$ whose integral is the desired position vector $\bar{X}_m$.

10.3 Unilateral and bilateral control

The systems described above strive to match the position of an end-effector to that of a workpiece or other target, and they rely on visual feedback to determine if that objective has been met. Motions demanded at the input to the machine controller produce related motions at the output. The coupling between the input and the output of the controller, whether mechanical, electrical or hydraulic, is one-way, the output being made to follow the input. Such systems are called *unilateral* tele-operators.

When a high degree of operational dexterity is required, as in assembly, maintenance or the manipulation of objects in space, then unilateral control has been found to be inefficient. For example, an operator could quite easily drive a unilateral system beyond its performance limitations. If the operator were to ask the end-effector to move more quickly than its actuators will allow, the resulting asynchronism between the control inputs and the manipulator movements could be confusing, leading perhaps to collisions. Since feedback is visual, the operator has no direct information about output forces, whether due to inertia or to reactions; this can lead to excessive demands being placed on the actuators, or, indeed excessive forces being applied to the workpiece. This can be avoided by the use of *bilateral* or force-reflecting tele-operators.

In a bilateral tele-operator there is a two-way coupling between the input and output of the controller: inertia and work forces exerted on the output can *back-drive* the input. Thus, in addition to the forward position loop which makes the end-effector follow the input demands, there is a backward loop to the input device from the end-effector. The servo parameter used for the backward loop may be position or force but, in either case, the operator senses all forces encountered during manipulation. Combined visual and force feedback give a better feel of what is actually going on, and this improves the accuracy and reduces the time for completion of a task.

Common error bilateral systems

Let us first look at those tele-operators in which both backward and forward loops use position information. The starting point will be that most

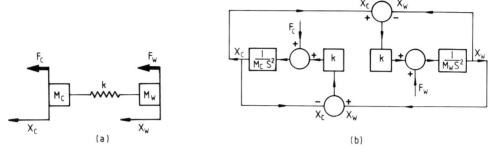

Figure 10.6 A simple example of a common-error bilateral system: (a) a model of a bar (b) block diagram illustrating two coupled systems driven by a common error.

elementary of tele-operators—a long bar for poking dangerous substances. Although a simple example, this is indeed a bilateral system, and it represents the basis of operation of the mechanical tele-operator of Figure 10.3. Either end can drive the other, and forces at either end are reflected to the other. This point can be illustrated by modelling a flexible bar as two masses M_c and M_w connected by a spring of stiffness k (Figure 10.6). External forces F_c and F_w are applied to the bar in the control space and in the work space. The equations of motion are:

$$M_c\ddot{x}_c = F_c + k(x_w - x_c) \qquad (10.1)$$
$$M_w\ddot{x}_w = F_w + k(x_c - x_w) \qquad (10.2)$$

Figure 10.6(b), which uses the Laplace operator s, shows these coupled equations in block diagram form. Two feedback loops can be identified. The right-hand one, the forward loop, driven by the error $(x_c - x_w)$, attempts to make x_w equal to x_c. The left-hand loop, the backward loop, is driven by the error $(x_w - x_c)$ and attempts to make x_c equal to x_w. Hence the spring, the controller in this case, attempts to make the displacement of each end equal to that of the other. This emphasizes the reversibility, or bilateral nature, of this simple system: either end can drive the other. Adding eqns (10.1) and (10.2) gives the relationship between the external forces:

$$F_c = M_c\ddot{x}_c + M_w\ddot{x}_w - F_w. \qquad (10.3)$$

This confirms the force-reflecting nature of the system. The operator's force F_c must overcome the inertia forces plus the force on the output end.

Mechanical systems such as those depicted in Figure 10.3 are simple to understand but difficult to build (Goertz, 1964). The ANL Model 1, conceived in the 1940s, developed over ten years to the Model 8 which became the workhorse of the nuclear industry. It was driven by a maze of tapes and pulleys, and had six degrees of freedom in addition to a hand closure control. But mechanical tele-operators have many problems that restrict their application. Backlash in gears, play in bearings and cable stretch combine to reduce positional accuracy, and the friction and inertia of

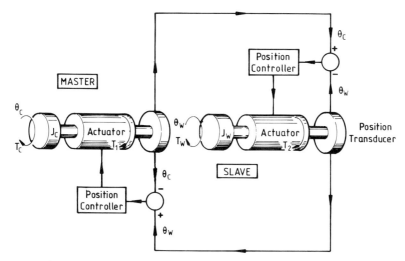

Figure 10.7 A bilateral electromechanical tele-operator.

the control arm can reduce force feedback to a meaningless level. In addition they cannot be used where distances between control and work spaces are large, such as in space, or where leakage across a barrier cannot be tolerated, such as in deep-sea applications.

Electrical transmission of information and power can surmount most of these problems, so let us examine the bilateral electromechanical system of Figure 10.7. Here the principle is the same: each actuator (rather than the spring of Figure 10.6) tries to equate the rotation at its end of the system to that at the other end. In order to achieve this a signal proportional to the rotation θ_c is fed forward to the work space, and θ_w is fed backward to the control space. Torque T_c is applied by the human operator and T_w is an external torque caused by gravity or by reaction.

Two position control loops, similar to those of Figure 10.6, can be identified. Neglecting energy dissipative terms, the control-side equations are

$$T_c + T_1 = J_c \ddot{\theta}_c \qquad (10.4)$$

where J_c is the inertia of the actuator and the input transducer. T_1, the torque generated by the actuator, is given by

$$T_1 = K_c(\theta_w - \theta_c) \qquad (10.5)$$

which assumes a simple proportional controller with gain K_c.

Hence, combining eqns (10.4) and (10.5) gives

$$J_c \ddot{\theta}_c = T_c + K_c(\theta_w - \theta_c) \qquad (10.6)$$

The similarity between this and eqn (10.1) is obvious.

Turning attention now to the workspace we can write

$$T_w + T_2 = J_w \ddot{\theta}_w \qquad (10.7)$$

and

$$T_2 = K_w(\theta_c - \theta_w) \qquad (10.8)$$

hence

$$J_w \ddot{\theta}_w = T_w + K_w(\theta_c - \theta_w) \qquad (10.9)$$

which is similar to eqn (10.2).

It is important that the displacement θ_w of the end-effector should follow that of the input transducer θ_c. Using the Laplace operator, eqn (10.9) can be rewritten as

$$\theta_w = \frac{K_w}{J_w s^2 + K_w} \theta_c + \frac{1}{J_w s^2 + K_w} T_w \qquad (10.10)$$

This indicates the presence of an undamped transient, which is not surprising since energy-dissipative terms have been ignored. If they were present the transient would die away to a steady state given by

$$\theta_{ws} = \theta_{cs} + T_{ws}/K_w \qquad (10.11)$$

This shows a torque-induced error, which can be reduced by increasing the controller gain.

Another important relationship is that between the torque T_c applied by the operator and the torque T_w acting on the end-effector. Multiplying eqn (10.6) by K_w and eqn (10.9) by K_c and adding gives

$$T_c = J_c \ddot{\theta}_c + (K_c/K_w)(J_w \ddot{\theta}_w - T_w) \qquad (10.12)$$

which is similar to eqn (10.3). Hence the operator feels the inertia torque at the control end, a proportion of the inertia torque at the working end and a proportion of the externally applied torque at the working end. In the steady state $T_{cs} = -(K_c/K_w)T_{ws}$. This is what we would expect: if an external torque attempts to increase the displacement of the end-effector then the operator will have to counteract that torque at the control end. If $K_c = K_w$ the torque feels identical to the operator.

The system is reversible: it can be *back-driven*. Expressions for θ_c and T_w can be determined by changing the suffices in eqns (10.10) and (10.12). Because both controllers use the error between θ_w and θ_c this particular form of bilateral tele-operator is sometimes called a *common error* system.

The earliest common error systems were built by Goertz's group at ANL in the early 1950s (Goertz *et al.*, 1953; Arzebaecher, 1960; Burnett, 1957). For example, the ANL Model 2 Manipulator used reversible AC motors for actuation. The common error principle was implemented by using the same current to activate both motors, with a simple circuit ensuring that this current was proportional to the positional error. Transients were damped by the use of rate signals from tachogenerators.

Force feedback bilateral systems

We mentioned earlier that some bilateral tele-operators use force instead of position information for the backward loop. Figure 10.8 shows such a system

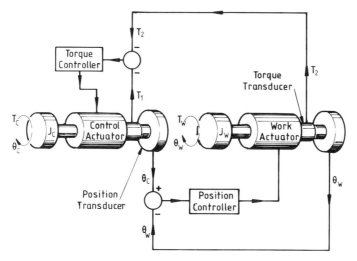

Figure 10.8 A force-feedback bilateral tele-operator.

in block diagram form. Torque transducers measure the torques developed by the control and work actuators, T_1 and T_2 respectively. As before, the forward position control loop attempts to match θ_w to θ_c. The backward torque control loop has the task of ensuring that the control actuator develops a torque T_1 which is the negative of, the reaction to, or the reflection of the torque T_2 generated by the work actuator. Hence the input to the torque loop is $-T_2$. This is an attempt to mimic the equal and opposite internal forces in the spring of the mechanical system of Figure 10.5.

The forward position loop is the same as that of the common error system of Figure 10.7, so eqn (10.10) can again be used to relate θ_w and θ_c. The torque equations are, as before,

$$T_c + T_1 = J_c \ddot{\theta}_c \tag{10.4}$$

$$T_w + T_2 = J_w \ddot{\theta}_w \tag{10.7}$$

and, assuming a simple proportional torque controller with gain K_t

$$T_1 = K_t(-T_2 - T_1) \tag{10.13}$$

or

$$T_1 = -K_t T_2 / (1 + K_t) \tag{10.14}$$

Substituting for T_1 and T_2 from eqns (10.4) and (10.7) gives

$$T_c = J_c \ddot{\theta}_c + \{ K_t / (1 + K_t) \}(J_w \ddot{\theta}_w - T_w) \tag{10.15}$$

This result should be compared with eqn (10.3) and eqn (10.12). In the steady state

$$T_{cs} = -\{ K_t / (1 + K_t) \} T_{ws} \tag{10.16}$$

Thus the system, like the common error bilateral tele-operator, allows the operator to feel the forces and torques at the end-effector.

Some of the earliest bilateral systems with force feedback were devised by Mosher (1967) and Mosher and Wendel (1960). They showed how the backward force loop could be implemented in an electrohydraulic tele-operator. The pressure difference across the piston of the hydraulic output actuator was measured by a transducer and the resultant electrical signal, being a measure of the output force, was fed back to a pressure-controlled servovalve which drove the control actuator. Since this type of servovalve provides an output pressure proportional to the applied current, then the reaction force on the control actuator was in turn proportional to the force supplied by the working actuator.

10.4   Analytical difficulties

The preceding section makes it clear that even the simplest single-degree-of-freedom tele-operator presents quite a complex analytical task. It is well-nigh impossible to produce a reasonably accurate mathematical model of a six-degree-of-freedom tele-operator. Chapter 4 illustrates some of the difficulties involved in modelling the dynamics of robot arms, but here the problems are compounded by the presence of the human operator who has non-linear time-varying characteristics.

Even in simple control tasks much effort has been devoted to finding an adequate mathematical model for the human operator (Kelley, 1968). In tracking, for example, where the operator uses a control stick to make a spot on a screen follow the position of a moving input spot, it has been shown that the human operator can be approximated by the transfer function

$$\frac{K(1 + T_2 s) \exp{(-T_1 s)}}{(1 + T_3 s)(1 + T_4 s)} \tag{10.17}$$

where K is the gain, adjusted by the human operator to the proximity of marginal stability; T_1 is the reaction time delay, ranging from 0.2 to 0.5 s for random stimuli; T_3 is the neuromuscular lag, normally between 0.1 and 0.16 s for the human arm; T_4 is the lag time constant—observed values are between 1 and 20 s in manual tracking, although it can have any value, depending on the system; T_2 is the lead time constant; this relates to the operator's ability to anticipate the movements of the input. It has been shown that the ratio T_2/T_4 in the operator's response will adapt to the optimum for a particular system. For example in position tracking, with a controlled element transfer function k, the ratio is high, whilst for acceleration tracking, with a controlled element transfer function k/s^2, the ratio is low.

Hence it is evident that any attempt at the analysis of a six-degree-of-freedom tele-operator can only be an approximation, giving at best some feel for the system and helping the designer to grasp the interrelations between the control parameters. These difficulties have led to a heavy reliance on the use of simulators for the design of tele-operators and other such systems that depend upon human controllers.

10.5 Control input devices

The *control input device* (CID) provides the actuating interface between the
human operator and the machine controller. It converts the operator's signal
(Figure 10.4) into a signal that can be understood by the machine: it is a
transducer.

On–off switches

The simplest CID is the on–off switch, used to control the power supply to
an actuator. These are common in rate control systems, the operation of the
switch causing the actuator to move at a fixed velocity. For full tele-operator
control, it is usual to mount these switches in a control box, arranged in an
array that bears some resemblance to the configuration of the manipulator.
A degree of anthropomorphism is often added by using three-position
switches so that a joint can be rotated clockwise by pushing the switch to the
right, rotated anticlockwise by pushing it to the left, and stopped by centralizing
the switch. Such control boxes are similar in some respects to those
described earlier for programming robots. Although simple, compact and
relatively cheap they have several serious shortcomings: only one actuator
can be driven at a time during precision manipulations; there is no force
feedback, restricting their use to unilateral teleoperators; control of velocity
requires potentiometers or multipole switches; operator identification with
the task is minimal.

Joysticks

The joystick is another commonly used CID. It is a moveable stick with as
many as six degrees of freedom. It is usual to mount control buttons and
switches on the stick within easy reach of the operator's fingers. Figure
10.9(a) shows how a joystick can be used to control the six degrees of

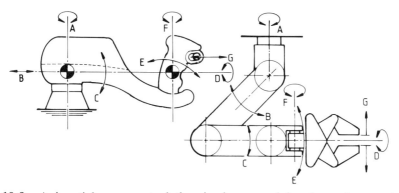

Figure 10.9 A joystick can control the six degrees of freedom of a manipulator,
and operate the end-effector.

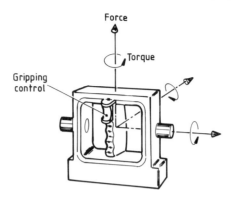

Figure 10.10   A six-degrees-of-freedom isometric hand controller.

freedom of a manipulator, and to operate the grasping action of the end-effector (Johnsen and Corliss, 1969). Signals from transducers on the various axes of the joystick are used to control the actuators on the corresponding axes of the manipulator. The transducers could be potentiometers if a position operated (isotonic) stick were used, or strain gauges if a force-operated (isometric or 'stiff stick') stick were used. In either case it is possible to control the manipulator in a proportional mode rather than the on–off mode associated with switch boxes. Joysticks are usually employed in rate-control systems, where the signals from the stick's transducers are used to control the velocities of the actuators. Operators find the joystick superior to the switch box, mainly because there is a more obvious correspondence between the operator's inputs and the manipulator's movements.

      Several new types of multi-degree-of-freedom hand controllers have been developed recently. One, intended for use in resolved motion rate control (RMRC), is shown in Figure 10.10. It is a six-degree-of-freedom isometric hand controller (Nevins and Whitney, 1977). The end-effector's velocities of translation along and rates of rotation about three axes are determined by the forces and torques applied about the axes of the hand controller. A trigger is also provided for opening and closing the grippers. A major advantage of such isometric controllers is their compactness, negligible movements at the controller allowing potentially unlimited movements of the end-effector.

Replica control

Another way of manoeuvring the end-effector to a desired position involves the use of a replica of the manipulator. The idea is simple. If each joint in the replica were fitted with a position transducer which provided an input signal to the corresponding joint servos on the real arm, then manual orientation of the replica would result in a similar orientation of the real arm. Such a replica was illustrated in the mechanical system of Figure 10.3, where cables ensured equal rotations of corresponding joints. This mechanical system demonstrates both geometrical similarity and geometrical identity for input and outputs, but this is not always necessary, and in many cases it is desirable to scale the output up or down with respect to the input. For example, remote manipulators for microsurgery and for the assembly of small parts require a reduction in scale, whilst tele-operators for handling large space structures require a magnification of scale.

Despite its cost and complexity, replica control, sometimes called *master–slave control* or *terminal control,* has proved to be attractive. To some extent it is like using a multi-jointed joystick, but the operator has only to concentrate on one motion—that of the replica hand. By moving it in the operator's set of world coordinates, the position transducers and the joint servos ensure that the real hand will move a proportional distance in its world coordinates. Another advantage is the easy accommodation of force feedback, allowing the full potential of the bilateral system to be exploited.

An increasing degree of anthropomorphism is achieved if the operator 'wears' the replica. The Handyman electrohydraulic bilateral tele-operator was built in the 1950s by General Electric, as part of the Aircraft Nuclear Propulsion programme (Mosher, 1967). Handyman had two arms and ten controlled joints in each arm/hand combination. The human controller stood inside or wore the master controls, so that each arm and hand movement was replicated at the working end of the teleoperator.

A maximum degree of anthropomorphism is seen in the so-called man-amplifiers (Figure 10.2(b)), in which the operator wears both the master and slave; these exoskeletal structures will be examined in more detail later in the chapter.

Voice inputs

All of the above CIDs require the operator to move or to apply forces by means of the limbs. Voice input gives a more direct form of control, offering the many advantages of natural language communication. A particularly attractive feature is the possibility of physical separation and physical mobility of operator and machine; a major disadvantage is the loss of the kinesthetic feedback associated with bilateral feedback.

Automatic speech recognition systems (ASR) are used nowadays for voice entry to data files and to high-security areas: they can help the handicapped: they can be used for learning aids and as toys. The ultimate aim is to develop an ASR that can recognize continuous speech uttered by any individual, but this is still a long way off and to date the most common systems are those that deal with isolated words uttered by a known speaker (*speaker-dependent* systems). Basically such systems include a series of data-processing stages that convert analogue acoustic signals to digital electrical signals, extract measurable features from these signals and match this extracted information with stored models.

Isolated word recognition systems (IWR) require that each word be bracketed by a brief silence (Baker, 1981). Speaker-dependent systems involve a training period during which the operator speaks a word several times in order to allow the establishment of a representative 'template' in computer store. The acoustic input is first digitized by sampling at 6 kHz to 20 kHz, depending on the desired quality, with each sample being quantized typically to 12 bits. Template construction begins by finding the start and finish of each word. The word is then divided into about 12 to 16 time slices or frames, and each of these is described or characterized by up to 32 acoustic parameters, including spectral features. These procedures are repeated in the recognition phase, the unknown word being divided into frames and the characteristics of the frames being determined. A comparison is then made between this set of characteristics and those of the template stored in memory, the computer calculating a score for each comparison. From these the best score is assumed to identify the input word.

Because of limitations on the size of store, most commercial IWRs restrict the number of templates to around 60 words of duration up to 2 s. This is more than sufficient to drive a tele-operator, since it is possible to achieve reasonable control of any degree of freedom using only five commands, such as *left fast, left slow, right fast, right slow* and *stop*. Sword and Park (1977) describe the use of such a system for unloading bombs from a rocket.

10.6   Control input devices for disabled people

The above discussion of CIDs was in the main concerned with industrial and military applications, but in fact tele-operators cover a much wider field, including orthotic and prosthetic devices for enhancing and extending the limited capabilities of the disabled. The wide variety of possible disabilities has required the development of many different specialized CIDs. Figure 10.11 illustrates a few possibilities.

Figure 10.11(a) shows how arm flexions or shoulder shrugs can be used to drive a prosthetic arm by means of a Bowden cable. Body-powered prostheses are light and robust, and since they are simple mechanisms they are relatively inexpensive (Childress 1973; Fletcher and Leonard, 1955). In addition they supply the operator with an element of force feedback and, to a lesser degree, proprioceptive feedback. However, these arms are usually restricted to one or two degrees of freedom since there is only a small number of suitable attachment points. They can be tiring to use, since operation often requires exaggerated body movements.

Body-powered prostheses are satisfactory for low-level amputees, but with higher levels of amputation, such as shoulder disarticulation, the number of control sites available, and the amount of power that can be generated, may be insufficient for such devices. It is then necessary to resort to the use of externally-powered prostheses. Body movements, such as those illustrated in Figure 10.11(a), muscle bulges, or the movement of vestigial limbs of those suffering from congenital deformities, can be used to trip

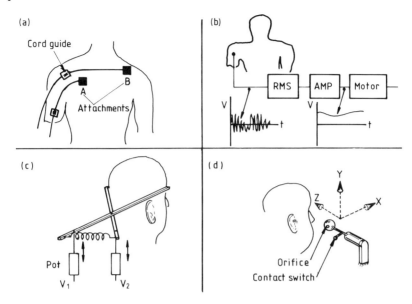

Figure 10.11 Control input devices for disabled people: (a) Bowden cables (b) electromyographic signals (c) head movements (d) mouth operation.

switches to operate the driving actuators. But another way of deriving the control inputs is to use the very small electrical potentials generated by muscular activity. These electromyographic signals (e.m.g.) typically micro-millivolts, can be detected by needle electrodes which record individual muscle fibre signals, or more usually (and less painfully) by metal disc electrodes placed on the skin to record a summation of the activity in many muscle fibres. Silver/silver chloride electrodes are commonly used for this purpose. The surface-recorded e.m.g. signal has to be processed in order to extract useful information. As shown in Figure 10.11(b) the signal appears to be random and noisy, but it has been shown that the gross force developed by the muscle is proportional to the root mean square of the signal, or to the average of the integral of the rectified signal (Patla *et al.*, 1982). Thus varying tensions in a muscle lead to varying voltages which can be used to control externally-powered actuators.

In extreme cases of disability, such as tetraplegics who have no control of legs or arms, it is necessary to look for other sources of body movement for control purposes. Fortunately many tetraplegics still have good control of head movements when the trunk is properly stabilized. Figure 10.11(c) shows a transducer that can measure head movements in the sagittal and frontal planes, i.e. rotations forward/back and right/left (Guitte *et al.*, 1979). The link with the head is realized by means of a V-shaped spring-loaded element, which preserves the cosmesis of the subject's frontal appearance. Head movements are also used occasionally to augment tele-operator control by people who are not handicapped—in particular to direct television cameras at targets (Charles and Vertut, 1977).

Another form of head-operated CID is illustrated in Figure 10.11(d). The joystick, shown on the right of the diagram, is mounted rigidly beside the operator's head and operated by mouth (Schmalenbach *et al.*, 1978). The axis lies across the front of the face, like a flute, so that the view is not obstructed. The joystick can be moved in all three spatial directions to produce three analogue signals with negative and positive polarity. Discrete signals can be obtained from tongue-operated touch contacts, or by blowing or sucking the small hole in the mouthpiece. Eye movements, voice inputs and e.m.g.s can also provide useful discrete signals.

10.7 Applications of tele-operators

We confine our attention here to three major application areas: space, deep sea and cybernetic anthropomorphic machines.

Space applications

The first faltering steps towards the use of tele-operators in space were taken nearly 20 years ago when NASA soft-landed the unmanned Surveyor on the Moon. TV pictures were relayed to Earth, and sampling of the lunar surface was carried out under remote control. The sampling arm had only three degrees of freedom—rotation in azimuth, rotation in elevation and extension in a radial direction. The operator could drive one motor at a time by digital commands from NASA's Goldstone Deep Space Network Station in California. More recently Viking 2 carried out similar operations on Mars. Attention has also been focused on the use of mobile tele-operators on the surfaces of the planets. They can collect and handle samples and carry out experiments, and they can be used for the repair and maintenance of equipment based on the planet.

The lunar and Martian explorations highlighted the difficulties of controlling tele-operators over extremely long distances. Even with radio waves moving at 3×10^8 m/s the delay between initiating an action on Earth and seeing its effects at these great distances is appreciable: 2.6 s for the Moon and from 6 to 44 minutes for Mars, depending upon whether it is at inferior or superior conjunction with the Earth. These delays pose considerable control problems, one possible solution being to control the tele-operators from orbiting spacecraft.

Another space application is illustrated in Figure 10.12(a) which shows a space shuttle (Orbiter) using its shuttle-attached tele-operator (SATO) to retrieve a free-flying tele-operator (FFTO). Such systems are being developed for placement, capture, maintenance and repair of satellites, for the correction of orbits, for the destruction of military satellites, for the construction of space stations and so on. (Onega and Clingman, 1972; Nathan and Flatau, 1978; Taylor, 1978).

The SATO can work independently to capture and deliver payloads,

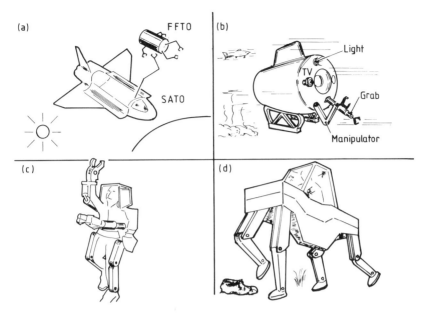

Figure 10.12 Application of tele-operators: retrieving a free flying teleoperator by a shuttle-attached teleoperator (b) a manned submersible (c) Hardiman, an exoskeletal machine (d) a walking truck.

but, from an energy viewpoint, it can be more cost-effective to station the Orbiter in a low Earth orbit and send the FFTO up to the higher geosynchronous orbits where most satellites are to be found. The arm of a SATO is typically 20 m long and can have seven degrees of freedom, the additional freedom allowing independent elbow positioning for hazard avoidance. The unmanned FFTO is controlled from the shuttle, and its use retains the advantage of a manned mission but eliminates the disadvantages of providing remote life support and hazard protection systems. Note that the FFTO in the illustration has four arms—two for grabbing and two for working.

Oceanographic applications

The past 20 years has seen a rapid growth in the exploitation of the ocean floor. Much of the work down to 200 m has been carried out by divers but, since most of the ocean floor is at least 3 km deep, the advantages of dexterity and flexibility offered by the human diver are being outweighed by a growing number of disadvantages (Datta and Kuo, 1978). Divers exposed to the open sea experience extremes of cold and pressure; the latter presents a particularly difficult problem. In water deeper than about 20 m the diver's internal body pressure has to be increased to match external hydrostatic pressure. This brings attendant hazards: too rapid compression can give rise to 'high pressure nervous syndrome': too rapid decompression can bring on an attack of the 'bends'. The supply of a suitable mixture of gases for breathing also has its complications. Air is satisfactory down to 50 m, but greater depths require the use of heliox. The percentage of oxygen has to be controlled, carbon dioxide has to be removed, nitrogen narcosis has to be avoided and a 'helium unscrambler' may have to be used to cancel the effects of voice distortion.

These problems can be circumvented by the use of manned submersibles (Busby, 1976), whether free-swimming or tethered to a mother ship (Figure 10.12(b)). Such devices allow the operator to remain in the relative safety and comfort of a sealed chamber maintained at normal room temperature and pressure. Indeed in some cases it may not be necessary or desirable to have the operator close to the workplace, and unmanned submersibles remotely controlled from a mother ship would be preferable (Charles and Vertut, 1977). This eliminates the need for a strong hull and a life-support sytem.

With the tele-operator arms attached to the submersible the operator is able to perform a variety of tasks. As in space applications it is necessary, because of reaction forces and torques, to use a grab to anchor the submersible to the workpiece. In addition, because of poor visibility at great depths, and in particular applications, it is desirable to take advantage of the force-reflecting properties of bilateral operators (Hotta *et al.*, 1977).

Applications for underwater tele-operators fall into three major categories: scientific, commercial and military (Johnsen and Corliss, 1969). Scientific investigations include the study of biology, botany and geology. Use of manipulators for selective sampling of rocks, shells and mud is to be preferred to the hit-and-miss nature of dredging. Commercial applications are now dominated by the construction and maintenance of offshore oil-wells, but there are many other areas where submarine tele-operators could be of use: salvage, photography, shipbuilding and repair and harbour maintenance, for example. On the military front there has been a need to recover practice and prototype ordnance; indeed the famous H-bomb recovery off the coast of Spain in 1966 encouraged a greater amount of research into underwater tele-operators. Another envisaged application is the use of tele-operators to instal and maintain ocean-floor military stations.

Cybernetic anthropomorphic machines

Engineers at the General Electric Company proposed this imposing name for that class of tele-operator in which operator and machine were so closely interlinked as to become practically one and the same thing. Two examples, both developed by General Electric, are illustrated in Figures 10.12(c) and (d). Hardiman an exoskeletal machine or man-amplifier (GEC, 1968), was constructed in 1966 for military applications: the 'walking truck' (Lefer, 1970), also built with military application in mind, took its first steps in 1968.

The operator of Hardiman stood inside an anthropomorphic structure built in two halves jointed together at the hips. Each half had fifteen degrees of freedom. The exoskeleton was parallel to the human structure everywhere except at the forearms, where the operator's arms were collinear with the exoskeleton's forearms. Power was supplied from a hydraulic system operating at a pressure of 200 bar. Bilateral electrohydraulic servos were used with force ratios between master and slaves of about 25. With this amplification the machine was able to lift and manipulate a mass of around 1000 kg, and to transport it through a distance of 8 m in 10 s.

Exoskeletal machines are potentially hazardous to the human they envelop. What would happen if power failed and the structure collapsed? What would happen if the machine became unstable? These problems received a great deal of attention from the General Electric designers, and their solutions included automatic locking of actuators if pressure failed, and physical linking of limbs to preclude the build-up of small errors between master and slave.

The work on Hardiman contributed considerably to the walking truck programme which commenced around the same time. In Chapter 3 the several advantages of legged vehicles vis-à-vis wheeled vehicles were discussed, and it was shown how a robot could be made to walk with a programmed gait. In the present example, the advantages of tele-operator control and of legged vehicles were combined. The result was a machine weighing 1500 kg, 4 m long and 3.6 m high with legs 2 m long, capable of moving at about 1 m/s on level ground. It was powered by a 70 kW petrol engine.

The machine followed the movements of the operator, whose arms and legs drove replica controllers. The operator's arms controlled the machine's front legs and his legs its rear legs. Bilateral hydromechanical servos, operating at 200 bar, were used to multiply the operator's forces by about 120 and movements by about 4. The operator has to control simultaneously 12 position servos and respond to force inputs from 12 actuators in each of the machine's legs. The intention was that, with the close interrelationship between operator and machine, the operator would only have to go through a simple crawling motion to make the quadruped move along amplifying and extending his every movement. The machine demonstrated an ability to walk forwards and backwards, to balance on diagonal legs, to climb over obstacles, to lift 200 kg masses and to turn around. In spite of this the project was not pursued beyond the prototype stage, for reasons based, one suspects, on the difficulty of the control problems presented to machine and operator alike.

10.8   Tele-operators for the disabled

In this section we examine two important areas of application of tele-operators: prosthetic limbs for the replacement of those missing because of congenital defects, or lost because of accident or violence, and (ii) remotely-controlled manipulators for the rehabilitation of paralysed people.

Artificial limbs

The design and manufacture of artificial limbs presents a difficult challenge to the engineer. Weight, reliability, actuation, power consumption, ease of control, cosmesis and cost are only a few of the factors that have to be considered (Jacobsen *et al.*, 1982; Peizer *et al.*, 1969). Weight constraints, for example, are at least as demanding as those found in aircraft design, for although the natural arm has a mass of about 4 kg, a prosthetic arm with mass greater than 1.5 kg would be unacceptable to an amputee.

The choice of power source is another area of debate. We saw earlier that power from body movement has its attractions, but its use is restricted to the simplest of devices. Electric batteries give the lightest and most compact form of external power source, but these advantages can be offset by the relatively high mass of the motor and reduction gearbox combination required for each degree of freedom. Pneumatics, offering low mass and high response, is the other main contender; the power in this case comes from bottles of pressurized liquified gas, usually carbon dioxide. Whether batteries or bottles are used, the problem of rundown and recharging has to be faced. Table 10.1 compares the performance of pneumatic and electric versions of the Otto Bock hand prosthesis (Schmidl, 1977).

Table 10.1

	Pneumatic	Electric
Pressure/voltage	5 bar	12 V
Maximum opening	65 mm	100 mm
Maximum grip pressure	7 kN/m²	15 kN/m²
Speed	45 mm/s	80 mm/s
Mass	340 g	450 g
Number of grips per charge	1300	4200

The pneumatic power pack consisted of 48 g of carbon dioxide at 5 bar, in a bottle of mass 350 g, and with dimensions 36×140 mm. The electrical power came from a 12 V rechargeable nickel–cadmium battery with a 450 mAh capacity, a mass of 280 g and dimensions $16 \times 58 \times 150$ mm.

Ease of control is another factor worthy of discussion. Much research has gone into the use of e.m.g. signals for the control of artificial limbs. A major problem in such applications is the difficulty of obtaining enough independent control signals from the body to control all of the required output functions. There have been two major attacks on the problem. In the first, the amplitude of the processed e.m.g. signal is used to define several output states by the use of threshold circuits. For example, a high-level signal from one site could be used to drive an elbow actuator while a low-level signal from the same site, might drive a wrist actuator (During and Mittenburg, 1967). An impressive application of this technique is seen in the hand prostheses developed by the Control Group at Southampton University (Swain and Nightingale, 1981). This requires the subtraction of processed e.m.g. signals from antagonistic muscles such as the flexor and extensor muscles of the forearm, or, for above-elbow amputees, the triceps and biceps. By selecting different levels of this control signal the patient was able to give the commands *grip, squeeze, manoeuvre, release* and *flat hand*.

Another method of controlling multiple outputs uses pattern recognition techniques. Most amputees retain a cortical image of the lost limb—the phenomenon of phantom limb perception. The amputee, imagining movements of the limb, sends signals to the remaining muscles or parts of muscles that would have been used for the imagined movements of the normal limb. For example, a below-elbow amputee attempting to raise his phantom hand contracts the remaining muscles in the forearm stump. It has been shown that different required outputs, such as finger flexion and finger extension, generate different patterns of processed e.m.g. signals for these muscles. This is illustrated in Figure 10.13 which shows the pattern generated by signals from six control sites when a patient demands six different actions from a phantom hand (Johnsson *et al.*, 1981). For finger flexion (FF), reading from top to bottom, the pattern could be described as 100011, and for finger extension (FE) 011000. Such patterns can be used for classification and identification of the desired hand movements. With n control sites it is theoretically possible to call up 2^n outputs. Control is natural, the operator not having to give any thought to the control inputs.

Particularly difficult problems are presented by high-level amputees who have lost most of the arm. Such people require shoulder–arm prostheses, and since most of the muscle groups connected with the movement of the arm have been lost, it is difficult to get direct control signals to drive each degree of freedom of the prosthesis. An interesting solution to this problem is given by Swain and Nightingale (1981), where the patient drives the hand in world coordinates by shoulder and body movements. Elevation of the collar bone raises the hand (Z), forward and backward movements cause the

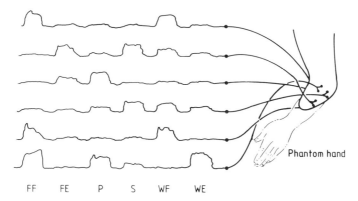

FF   FE   P   S   WF   WE

Figure 10.13 Pattern recognition for controlling multiple outputs: FF = finger flexion, FE = finger extension, P = pronation S = supination, WF = wrist flexion, WE = wrist extension.

hand to follow spatially (X) and rotation of the upper body left/right causes the hand to follow in the Y-direction. These body movements are detected by potentiometers mounted on a body harness. The necessary transformations from world to machine coordinates are carried out by a micro-processor.

Tele-operators for tetraplegics

One of the basic purposes of tele-operators is to allow people to gain access to regions which would normally be inaccessible. For tetraplegics this inaccessible region is the immediate environment. Two main techniques have been developed to assist such people: *orthoses* and *teletheses*. Orthoses, in the form of exoskeletal bracing and externally powered actuators, have been used in rehabilitation engineering, and computer-controlled powered orthoses have moved paralysed arms to perform desired manipulations (Correll and Wijnschenk, 1964). But the construction of these devices is difficult, and since most paralysed limbs are insensitive there is a danger that the disabled person's arm could be inadvertently injured by the orthosis or by the environment.

Telethesis, on the other hand, offers a useful and practical aid for tetraplegics. Telethesis is the name given (Guitte *et al.*, 1977) to that class of tele-operator systems for the disabled which, unlike prostheses and orthoses, are not directly attached to the operator's body. Figure 10.14(a) shows a telethesis in the form of a manipulator arm driven by the movements of a disabled operator's head. It is important that such a telethesis be situated in an ordered environment—an environment adapted to the dexterity and versatility of the manipulative system (McGovern, 1977). The disabled person is seated in the centre of this environment, where he can overlook the complete working range of the manipulator. The working range should be provided with borders to prevent objects from getting out of reach. Storage spaces must be provided, and if these are beyond the reach of the manipulator then it is necessary to arrange for some form of automated

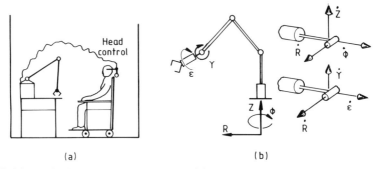

(a)   (b)

Figure 10.14 Telethesis for tetraplegics: (a) use of head movements (b) a resolved motion-rate controller using three inputs to drive five machine coordinates.

system of moving belts, trays or shelves to relay objects to and from the working space.

A major difficulty lies in the provision of a sufficient number of input signals. Figure 10.14(b) shows how a five-degree-of-freedom jointed arm manipulator can be driven by a three-degree-of-freedom head-operated joystick of the type described earlier (Schmalenbach *et al.*, 1978). Operating as an RMRC, it requires two control modes to be defined: one for gross movements to position the gripper, when the CID controls Z, ϕ and R, and one for orienting and fine positioning of the gripper when γ, ε and R are controlled. Discrete signals for switching between modes and for operating the gripper are obtained from tongue-operated switches, or by blowing or sucking through the small hole in the mouthpiece. This system allowed its disabled operator to perform a variety of tasks in a relatively short time. For example, it was possible to operate a motor-driven rotary shelf using a switch, take a book from the shelf, lay it down and open it—all in 1.5 minutes.

Rate control appears to be the best form of control of a telethesis. In cases of emergency, releasing of the stick brings the manipulator to a quick halt; and a manipulator position, reached perhaps after considerable control effort by the operator, can be maintained if the operator wishes to take a short rest. In McGovern (1977) the analogue input from the control stick was divided into three ranges. The first, roughly the first 10% of stick movement, was reserved for lost motion and avoided unintended motions of the manipulator such as those caused by accidental impacts on the stick. In the second range, another 10%, the actuators were driven very slowly, a useful feature during the learning process and for precise positioning. In the third range, velocity was proportional to stick displacement.

With regard to safety it was found (Guitte *et al.*, 1979) that the relationship between head movement and forward/backward movements of the gripper was important. Ergonomically it would appear that the most satisfactory arrangement would be one in which the gripper moved in the same direction as the head. However, this could lead to an unstable system. For example if contact were established between the gripper, or something carried by it, and the head, then withdrawal of the head would result in the head being pushed further backwards. This can be avoided by reversing the spatial correspondence between the forward/backward movements of head and gripper, at the cost of more extensive training of the operator.

10.9 Computer augmentation

Some of the tele-operators described above used computers to assist the human operator. In some cases complicated transformations from world coordinates to machine coordinates were delegated to the computer, greatly reducing the difficulty of the human operator's task. Computer augmentation of tele-operator systems offers many advantages. Extremely complex tasks (Sword and Park, 1977) can be performed by combining the human ability to plan, to interpret and to react to the unforeseen, with the computer's ability to detect, to store, to process and to repeat multichannel signals.

These characteristic differences in the attributes of people and machines suggest a multilevel control structure with varying degrees of human and machine participation at the different levels. Three main levels have been identified (McGovern, 1977): manual control, augmented control and artificial intelligence. With manual control, such as occurs in the traditional master/slave tele-operators, the human operator carries out all the tasks of signal measurement, processing and control, and provides all system commands. At the other extreme, that of artificial intelligence, the system is entirely under computer control and acts as an autonomous robot.

Between these extremes lies a continuum of control schemes in which the machine augments the human operator to varying degrees. There are two distinct areas of augmented control—supervisory control and automated control. In the supervisory mode, control switches intermittently between operator and computer, whilst in automated control the human operator and the computer work together, each being responsible for a separate function.

An experiment related to supervisory control is described in McGovern (1977) where the task was to grasp and manipulate blocks. The human operator's task was to move the gripper until it contacted a block. Once this was accomplished, control switched from operator to computer, and a routine for grasping the block was executed. When the block had been grasped control was then returned to the human operator. A practical example of a similar control scheme involved the location and acquisition of small bombs from a rocket nose section (Sword and Park, 1977). In this case the operator positioned the end effector over the bomb using voice inputs and, in response to the command 'grab it', a PDP-10 computer executed a series of routines which led to the bomb being grasped. Again, in Guitte *et al.* (1979), one finds a description of a telethesis for tetraplegics in which the disabled person's task of manipulation is simplified by employing a grasping reflex.

In the other type of augmented control, automated control, the human operator and the machine work together (Bejczy, 1980). The use of computers to work out complex transformations is one example where the task of manipulation is such a dual responsibility. Another example is the 'soft touch' ability that allows some grippers to apply just sufficient force to prevent objects slipping from their grasp (Guitte *et al.*, 1979). The operator exercises no control over this function: it is delegated to the computer.

Augmented control is a very active research topic at present. What is the optimum amount of augmentation for a given task? Which of the human supervisor roles are the most important: planning, teaching, monitoring, intervening?

10.10   Recapitulation

Robots are automatic, but tele-operators rely on humans for control. They extend human capabilities across barriers of distance and environment, and in the form of prosthetic and orthotic devices and teletheses they can assist disabled people.

Rate and position control strategies have been compared. Bilateral tele-operators, with their ability to be back-driven, were shown to be superior to the unilateral type: they provide the force feedback which is essential for accurate control. Bilateral control exists in two major forms; common error systems and force feedback systems.

A variety of control input devices has been discussed, including switches, joysticks, replica controls and voice inputs. For disabled people movements of head, mouth or vestigial limbs can be used for control purposes. Voice and electromyographic signals have also been used.

Several applications have been described. It was argued that some form of computer augmentation is likely to be necessary for all but the simplest of tasks.

Chapter 11

Economic and financial factors

11.1 Introduction

We now turn our attention to the economic and social aspects of robots. The world population of robots is analysed, and the new industry which has grown up out of the demand for robots is examined.

For a potential user, the costs involved will have a strong influence on whether or not to install a robot. The authors have therefore considered it important to examine the methods of financial assessment.

11.2 The robot population

The world's population of robots has been growing rapidly since the first one was installed by Unimation in a forge in 1962. Many statistics are published every year on the numbers of robots in the world and their applications. One of the difficulties encountered in comparing statistics from different countries is the variation in definition of the machines to which they refer. For example, a machine that would be called hard automation in the United Kingdom or in the USA might be called a robot in Japan. The survey conducted by the British Robot Association at the end of each year (BRA, 1985) and published in February of the next year is particularly useful, since the differences in definitions are accounted for, and the figures compare like with like. The figures from this survey showing the numbers of robots in different countries were quoted in Chapter 1, and are illustrated again here in Figure 11.1(a). It is interesting to see how these population figures relate to the size of each country's industrial base, and Figure 11.1 shows, for the

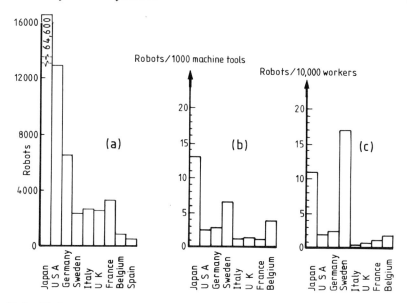

Figure 11.1 Robot population statistics 1984: (a) total in each country (b) number of robots related to number of machine tools (c) number of robots related to working population.

eight countries with the largest robot populations, the ratio of robots to workers and to machine tools. Japan and Sweden stand out on both indicators as the most automated countries.

The factors contributing to the extent of automation in a country have as much to do with the types of manufacturing industry prevailing there as with the general level of mechanization in the industry of that country (Gerstenfeld and Berger, 1982); Figure 11.2 illustrates this point by

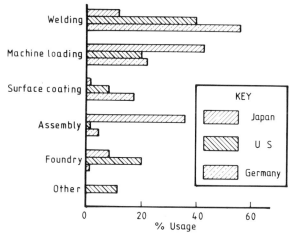

Figure 11.2 Percentage usage of robots in Japan, USA and West Germany.

comparing the percentages of robots in Japan, USA and Germany used on different applications. The USA figures provide a convenient benchmark, since they represent a broad industrial base. The automated industries (from motorbikes through cars to larger industrial vehicles) predominate in Germany, giving rise to heavy usages for welding (e.g. spot-welding for car bodies) and surface-coating (e.g. painting and undersealing of car bodies). The fact that many robots in Japan are employed on machine loading and assembly tasks reflects the concentration there of plants producing machine tools (including robots themselves) and consumer goods.

Manufacturing is often categorized into three areas: custom manufacturing or one-off production, batch production and mass production. The above statistics indicate that robots are finding applications in a range of manufacturing from medium batch to mass production. The sensory and artificial intelligence capabilities of robots are becoming more accessible to the commercial user, and it is expected that this will lead to a growing use of robots in small batch production, where sensor-based intelligence is needed to cope with the variations in work and environment associated with production in small numbers. The importance of this trend cannot be overemphasized: one survey (Ross, 1981) has indicated that production in batch sizes from 10 to 50 units accounts for more than 80% of all types of products manufactured in the world.

The previous discussion would indicate that the UK robot population has not yet reached its peak, and this is confirmed by Figure 11.3(a) (BRA, 1984) which shows the rise in the UK robot population over recent years. It is interesting to note that the population of robots in the UK, viewed from the standpoint of the number of different applications or number of different sites, is thinly spread. This indicates that many robot installations are first-time trials for the company and, if successful, could well lead to further

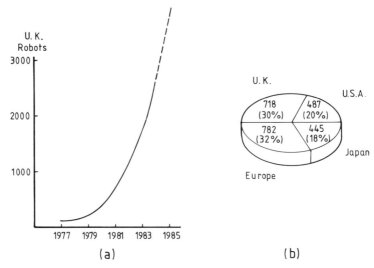

Figure 11.3 UK robot population: (a) growth (b) orogin of robots.

more ambitious projects, made easier by the fact that experience has been gained on initial installations.

Figure 11.3(b) shows the origin of all robots in the UK. Although only 30% of robots installed by the end of 1984 were manufactured in the UK, it is known that 33% of the robots installed during 1984 were locally made. More than half of the robot suppliers in the UK sell British-built machines, indicating a growth in the UK robot manufacturing industry. Western Europe countries, with the exception of Sweden and Germany, did not exhibit a concerted effort to automate production when the technology first became available in the 1960s; the recent upsurge in this area has been more of a reaction to the competittion from other parts of the world where the effects of automation were beginning to show themselves in lower production costs and higher quality. Thus the initial demand for robots in Western Europe has been met largely by imports from USA, Japan and Scandinavia, and this demand provides the encouragement for the growth of more locally-based robot manufacturers.

The figures for robotic applications in the UK were first mentioned in Chapter 1, and the numbers of robots in the main application areas are again illustrated here in Figure 11.4. The list confirms that the majority of robots are direct replacements for tasks previously done manually. The automotive industry (including not only vehicle manufacturers but all component manufacturers) has the tasks to which robots are easily applied and the capital to instigate large installations; hence the areas of spot-welding, surface-coating and injection-moulding, whilst not yet saturated, are well established. As sensor technology progresses, the arc-welding, assembly and inspection sectors can be expected to grow.

It was stated earlier that the UK's robot population was spread thinly, and this is confirmed by Figure 11.5 which illustrates some results from Northcott and Rogers (1984). The survey calculated, to the nearest 100, the number of robots in use in 1983 and the number expected to be in use in 1985, categorized by company size and type of industry. As might be

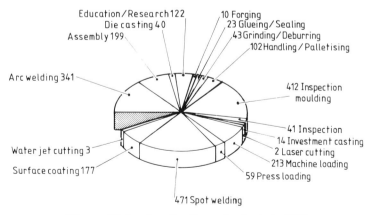

Figure 11.4 Application of robots in UK.

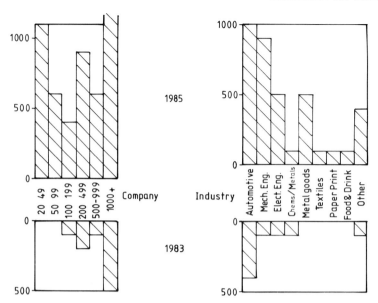

Figure 11.5 Actual (1983) and expected (1985) use of robots in UK categorized by company size and type of industry.

expected, smaller companies and the more traditional industries are not the first manufacturers to venture into robotics; however, the 1985 expectations indicate an awareness of the capabilities of robots, and the demand from the small companies for more flexible machines to handle small batch production provides the impetus for research and development. These comments apply equally to other areas of automation, and steady, though not quite so dramatic, population growths are observed in CAD work stations, CNC machine tools, programmable logic controllers and pick-and-place devices.

11.3   Robot industry

The modern robot industry has its roots in many technologies, notably manipulators (one was developed by Babbit in 1892 (US Patent No. 464 870)), servomechanisms, and computers. Patents in the names of Pollard in 1942 and Roselund in 1944 refer to paint-spraying machines which in fact performed the functions of a playback robot. George C. Devol, Jr. is credited with the invention of the robot, with his 1946 patents. From these inventions a worldwide industry has developed, involving not only the process of manufacturing robots themselves but also requiring a host of smaller companies for the supply of components (mechanical, electronic and software), the design and manufacture of add-on peripherals (e.g. purpose-built grippers, vision systems), and maintenance and other service functions.

Devol's early patents were sold to the Consolidated Diesel Corporation (CONDEC) which set up a subsidiary Unimation Inc. (from UNIversal autoMATION) to develop commercial machines, leading to the 1962 installation previously mentioned.

In 1974 the French car manufacturers Renault, having an interest in using robots in their own manufacturing operation, instigated a development project to produce machines for their own use. Renault subsequently became a manufacturer and supplier of industrial robots on the open market.

Also in 1974 appeared the first robot controlled by a microcomputer, the T^3 (the Tomorrow Tool Today), manufactured by the Cincinnati Milacron Corporation, a well-established maker of machine tools.

These three examples are good illustrations of the different origins of robot manufacturers. Many large manufacturers are relatively new companies, set up specifically to make robots. Others have emerged from potential robot users who, having developed robots, have found that they had machines which not only could be used inside the company but could be marketed outside, e.g. General Motors or IBM. In many respects a robot is just another machine tool, so the technology is a natural extension of the activities of established machine-tool manufacturers; the development of ASEA products from automatic welding equipment to welding robots and then to general-purpose robots therefore seems a logical process.

Figure 11.6 shows the growth of the market shares of the main manufacturers of robots in the world. A comprehensive summary of industrial robot systems available in the USA is to be found in the source of this chart (Hunt, 1983).

The state of the robot industry in the UK is often assessed by direct comparison with that in the USA and Japan. This can be misleading since

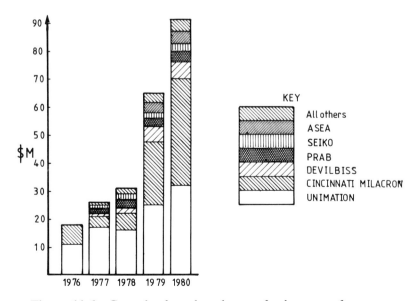

Figure 11.6 Growth of market shares of robot manufacturers.

the industries of these countries are sustained by a large home market. The UK robot industry compares better with those in France, Italy and Belgium, which are countries with similar markets to the UK (Figures 11.1 and 11.2).

It was remarked earlier that UK manufacturers were increasing their market share, the figure standing at 34% in 1983, but exact analysis is difficult because of the complicated nature of the industry. The three main manufacturers, ASEA, Cincinnati Milacron, and Unimation sell their models in the UK, but Unimation only undertakes some manufacture in the UK. Most Japanese robots are imported directly, e.g. Daros robots made by the Dainichi-Kiko group in Japan and imported by Dainichi-Sykes Robotics Ltd. Licensing deals will play a significant part in robot manufacturing; notable examples are GEC, whose agreement with Hitachi allows them to add to the range of robots inherited from Hall Automation, and the 600 Group, which has a manufacturing agreement with Fujitsu Fanuc. Comprehensive data on all robots available in the UK is to be found in the UK Robots Industry Directory (BRA, 1984); similar information for Europe and the USA can be found in Cugy (1984) and Hunt (1983), respectively.

11.4 Financial justification for robotics

Many industrialists (Yonemoto, 1981) consider the potential reduction of labour costs to be the most attractive feature of robots (see Chapter 1). Indeed the decade from 1970 to 1980 saw the hourly running cost of a robot fall below the hourly cost of manual labour in the industrialized nations (Lewis *et al.*, 1983), (Figure 11.7(a)), and this situation is expected to continue to favour the robot. If the primary objective is to reduce labour costs, then methods are needed for assessing whether the investment in an installation (the capital cost of the equipment) justifies the return (the savings due to reduced labour costs).

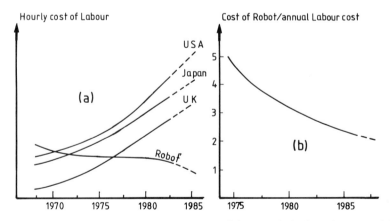

Figure 11.7 Labour and robot running costs: (a) growth in hourly cost of labour as robot running cost falls (b) fall in ratio of robot/labour cost.

Payback

Consider first the financial justification of the replacement of one manual worker by one robot. The simplest rule sets a permissible limit on investment, the limit being expressed in terms of the number of years of manual labour costs it would buy. This rule assumes that robot running costs are negligible (Figure 11.7(a) showed some justification for this). As wages rise at, say 7% per annum, and the average price of a robot does not change Figure 11.7(b) (Kuwahara, 1982) shows that direct robot replacement for manual worker installations will become more attractive on this basis.

In reality, in addition to the cost of the robot, its installation takes time to plan and commission. This construction stage incurs expenses, but once the robot is in operation these will be gradually offset by the resultant savings. The project will break even at the end of the payback period. The majority of current robotic installations exhibit payback periods of between one and four years: Figure 11.8 shows the range of payback periods of recent robotic applications (Lewis *et al.*, 1983).

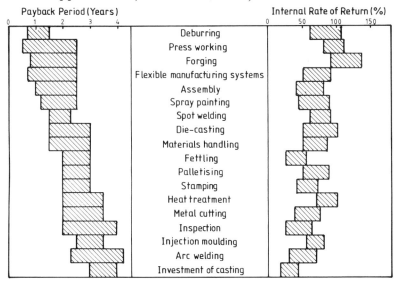

Figure 11.8 Payback period and internal rate of return for robotic applications.

Unit costs

When investigating a proposed automatic manufacturing system it is important to know how costs vary with the output rate of the system. Figure 11.9 gives an idea of trends. It shows that the cost of producing one unit manually tends to be constant since one worker can only produce a fixed number of units in a given time. On the other hand, within the limits of the output of the system, unit costs for hard automation reduce linearly with

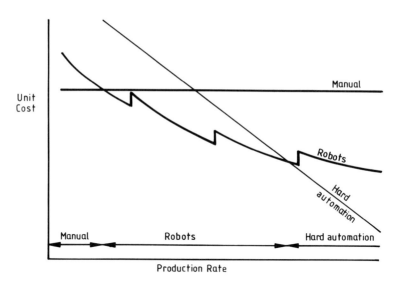

Figure 11.9 Variation of unit cost with production rate for manual, flexible and dedicated systems.

increasing output rate. The more units a given machine can produce, the better.

Because of the capital cost of equipment, robot systems will have higher unit costs for low output rates than will manual systems. Conversely for the same reason, they will have lower unit costs than hard automation. At the higher output end robots cannot compete with the mass production capabilities of hard automation. Hence Figure 11.9 shows that robot systems find their most efficient applications in medium-rate production systems (Scott and Husband, 1983).

To be precise the unit cost is the sum of contributions from fixed costs, which are independent of output rate, and variable costs which depend on the number of units produced. Fixed costs include depreciation and the proportion of fixed overheads assigned to the robot. The annual depreciation of equipment arises from the difference between its capital cost and scrap value spread over its useful life. Variable costs include the cost of material, labour and running cost of the machinery.

A product is manufactured manually at an annual rate of 480 000. Would an automatic assembly system be more efficient, and if so which type? Owen (1982a), cites the following example which illustrates the calculations:

Existing manual system

4 workers at £4000 p.a. each single shifts.
5% rejects
1 inspection worker at £6000 p.a.
Cycle time 15 s

Proposed flexible automation system

Capital cost £40 000
Depreciation over 5 years to zero value
Cycle time 10 s

Proposed hard automation system

Capital cost £100 000
Depreciation over 5 years to zero value
Cycle time 1 s

	Manual	**Hard automation**	**Flexible automation**
	£	£	£
System cost	500	100 000	40 000
Depreciation		20 000	8 000
Fixed cost	7000	27 000	15 000
Variable costs	3	2.95	2.95
Sales price	5	5	5
Break even at	3500	13 170	7 317
Production cost	£1 447 000	£1 443 000	£1 431 000
Cost reduction from manual		4 000	16 000
Payback period		25 years	$2\frac{1}{2}$ years
Unit cost	£3.0145	£3.0062	£2.9812

The flexible automation system is justified on a unit cost basis. The calculations do not assume any savings due to reduced reject rates or the possibility of dispensing with the inspection activities; these assumptions would be found to reduce the payback period to 1.09 years.

Discounted cash flow

When analysing the cash flow from a project over a long period, it is necessary to take into account the fact that an amount of money received at some future date is not worth as much as the same amount received now; basically because money, available at present, could be invested and could therefore accrue interest. The discounted cash flow technique allows future cash flows to be weighted accordingly so that they may be directly compared to present-day values. It is only possible to use predicted interest rates, but the technique is important for two reasons. First, it allows a genuine comparison between different installations where cost savings may occur over widely varying time scales, and secondly any capital project must at least pass the basic accountancy test of being able to generate a better return than if the money were simply invested.

Consider a project which requires an initial capital outlay of £K and has an investment horizon of H years (the investment horizon would normally be the useful life of the equipment). If the project generated net cash flows of $C_1, C_2, \ldots, C_H$ in years $1, 2, \ldots, H$, the net value NV of the project, at zero interest rate is

$$-K + C_1 + C_2 + \cdots + C_H \tag{11.1}$$

The value of the cash flow for year N, related to the present time, termed the present value (PV), is given in terms of the annual interest rate (i) by

$$\frac{CN}{(1+i)^N} \tag{11.2}$$

Thus the net present value (NPV) of the project is

$$-K + \frac{C_1}{1+i} + \frac{C_2}{(1+i)^2} + \cdots + \frac{C_H}{(1+i)^H} \tag{11.3}$$

Setting the NPV equal to 0 and solving for i gives the internal rate of return (IRR) of the venture, which is the true interest earned up to the investment horizon from the initial investment K. Figure 11.8 shows the range of IRR experienced from different robotic applications.

Unit cost-based comparisons

The financial concepts introduced are now applied to the problem of choosing between different manufacturing systems. The decision is to be based on a comparison of unit costs for different proposed automatic systems (e.g. hard automation, robots) to that of an existing manual system. A thorough treatment of this problem is to be found in Gustavson (1983), on which this section is based.

The unit cost C of producing a product is the sum of fixed and variable costs:

$$C = C_F + C_V \tag{11.4}$$

where C_F is the fixed unit cost, and C_V is the variable unit cost.

The fixed unit cost is given by

$$C_F = P_A/Q_A \tag{11.5}$$

where P_A is the annualized installed system price, and Q_A is the annual production quantity.

The annualized installed system price is the total fixed cost to be charged to the system for each year of its useful life and is given by

$$P_A = Kf \tag{11.6}$$

where f is the annual cost factor.

For a proposed system the installed cost K is not known at the planning stage. It is often calculated as a multiple of the hardware cost, the multiplying factor ranging from 1.0 for manual systems through 1.5 for hard automation to 2.0–5.0 for flexible automation.

If a proportion V of the hardware cost is expected to be recoverable after H years, this recovered cost has a PV of

$$\frac{V}{(1+i)^H} K$$

The true PV of the installed cost is therefore

$$\left[1 - \frac{V}{(1+i)^H}\right] K$$

Since this money could have been invested at the minimum acceptable IRR (i) for the duration of the project, it is not sufficient to charge a simple fraction $1/H$ of the cost for depreciation each year; the cost must be weighted to take account of the IRR. The true annual depreciation charge is therefore fK, where the annual cost factor f is given by

$$f = \left[1 - \frac{V}{(1+i)^H}\right]\left[\frac{i(1+i)^H}{(1+i)^H - 1}\right] \tag{11.7}$$

The variable unit cost arises from labour and operating costs

$$C_V = \frac{(W \times L_H) + O_H}{Q_H} \tag{11.8}$$

where W is the number of operators in the system, L_H is the hourly labour rate, O_H is the hourly overhead rate and Q_H is the hourly production rate, which is related to the annual production quantity by

$$Q_A = S \times 8D \times Q_H \tag{11.9}$$

where S is the number of eight-hour shifts per day and D is the number of working days per year.

Equations (11.4), (11.5), (11.6), (11.8) and (11.9) combine to give the unit cost as

$$C = \frac{1}{Q_H}\left[\frac{Kf}{8SD} + (W \times L_H) + O_H\right] \tag{11.10}$$

where f was defined in eqn (11.7).

Equation (11.10) gives the unit cost in terms of the production rate for an existing or proposed system, and typically generates curves of the form shown in Figure 11.9: this allows selection of the most cost-effective system for a given production rate. The discontinuities in the curve for the robotic system occur when a robot cell is used to capacity and another cell has to be provided for further increases in production rate.

The value of this analysis may be illustrated by an example. An existing metal-cutting operation involves four machines served by two operators working on each of two shifts, the cell producing 30 products per hour. Three alternatives are available:

    (i) Keep existing system
   (ii) Introduce a robot, the cell only needing one operator each shift.
  (iii) Introduce a pick-and-place device, needing one operator, but slowing down production to two-thirds of the existing rate.

    The figures for each of the alternatives are respectively

    (i) Hardware cost 0
       number of operators $W = 4$
       overhead costs $O_H = £4/hr$
       labour rate $L_H = £5/hr$
       production rate $Q_H = 30/hr$
   (ii) Hardware cost £50 000
       number of operators $W = 2$
       overhead costs $O_H = £5/hr$
       labour rate $L_H = £5/hr$
       production rate $Q_H = 30/hr$
  (iii) Hardware cost £20 000
       number of operators $W = 3$
       overhead costs $O_H = £4.40/hr$
       labour rate $L_H = £5/hr$
       production rate $Q_H = 20/hr$

    The practice of eight-hour shifts for 250 working days per year is established, and in all cases the recoverable value is considered to be 5% of installed cost after five years, i.e. $V = 0.05$, $H = 5$. The minimum acceptable IRR is 10%, i.e. $i = 0.1$. The annual cost factor is therefore, from eqn (11.7), $f = 0.256$. If the installed cost K is considered to be double the hardware cost, then eqn (11.10) gives unit costs:

    (i) $C = \dfrac{1}{30}(0 + 4 \times 5 + 4) = £0.80$

   (ii) $C = \dfrac{1}{30}\left\{ \dfrac{100\,000 \times 0.256}{8 \times 2 \times 250} + (2 \times 5) + 5 \right\} = £0.71$

  (iii) $C = \dfrac{1}{20}\left\{ \dfrac{40\,000 \times 0.256}{8 \times 3 \times 250} + (3 \times 5) + 4.60 \right\} = £1.06$

    On a unit cost basis, the second alternative, the robot installation, is justified.

    The foregoing analysis does not take into account the advantages of the flexibility of robots and other reprogrammable equipment. Hard automation equipment may not be able to accommodate changes in product design, but flexible systems are easily reprogrammed, and new designs often only require some retooling.

Methods of quantifying the savings due to flexibility may be found in the literature (Airey and Young, 1983; Bryce, 1983; Dapiran and Manieri, 1983; Gerstenfeld and Berger, 1982; Owen, 1982b; Scott and Husband, 1983). In many cases an ability to deal with regular design changes is desirable, e.g. car-body spot-welding lines. And it is worth noting that the robots and bowl and magazine feeders involved in flexible automation systems can find a variety of uses in the factory. They are therefore of great value to the user.

11.5 Recapitulation

The robot population has been examined with regard to growth, spread of applications, usage by different industries and the effect on the robot industry itself. Any robot installation must be justified financially, and the payback period is an indicator for projects with high savings. When comparing manual, dedicated and flexible automated systems, unit cost calculations taking into account the time value of money is a rigorous method of assessing viability.

Chapter 12

Safety and social factors

12.1   Introduction

Having examined some of the economic aspects of robots, we now turn to a consideration of the social aspects. One of the more important of these concerns safety in the factory.

Although robot installations have a much better safety record than the installations which they replace, they still present some problems.

The inherent improvement in safety is a desirable feature of robots: so also is the wealth created on a global scale as a result of improved productivity. However, it would be foolish to ignore the unsettling changes in employment patterns brought about by automation.

12.2   Safety aspects

In comparison to other machine tools, robots have a good safety record; many installations make a major contribution to safety by taking over dangerous jobs from human operators. Indeed, Japan boasts of a grant scheme which encourages small- to medium-scaled companies to introduce robots for this purpose (Hasegawa and Sugimoto, 1982). However, we are concerned here with the problem of making robot installations safe for human operators, no matter what the driving force for their introduction. The good safety record of robots should not detract from the importance of this process, which can only be undertaken effectively at the initial planning stage of an installation.

None of the UK regulations deals specifically with industrial robot installations, but a publication from the Machine Tool Trades Association (MTTA, 1982) gives practical guidelines. These proposed regulations have been submitted to the International Standards Organization, along with draft regulations from Japan and Germany, and an international safety code for robot installation is expected soon.

The safety process has two stages: identifying the safety hazards and identifying systems to deal with the hazards.

Hazard analysis

With respect to safety, the unpredictability of robot actions is a fundamental difference between them and other machines. Hunt (1983) sites several mistaken assumptions which operators have made about robot actions:

- If the arm is not moving, they assume it is not going to move.
- If the arm is repeating one pattern of actions, they assume it will continue to repeat that pattern.
- If the arm is moving slowly, they assume it will continue to move slowly.
- If they tell the arm to move, they assume it will move the way they want it to.

An example of the second occurs in welding applications where a robot may bring the welding tip round to a cleaning station after a fixed number of welding cycles. Again, with respect to the third, operators may not appreciate that a large hydraulic robot, normally only moving at slow speeds, is capable of violent and erratic motion simply due to a foreign particle causing a servovalve to stick.

The sources of hazards have been identified (Percival, 1984) as:

- Control errors caused by faults within the control systems, errors in software or electrical interference.
- Unauthorized access to robot enclosures.
- Human errors when working close to a robot, particularly during programming, teaching and maintenance.
- Electrical, hydraulic and pneumatic faults.
- Mechanical hazards from parts or tools carried by the robot or by overloading, corrosion and fatigue.
- Hazards arising from the application, such as environmental hazards of dust, fumes, radiation.

Detailed analysis of types and sources of accidents may be found in the literature (Sugimoto and Kawaguchi, 1983), and apart from those arising from the application itself, the human faces two types of hazard from a robot: impact and trapping.

There is a possibility of impact with moving parts of the robot or with items being carried by it. The greater the speed involved, the greater the danger of flying objects in the form of parts released from the gripper during turning motion, or items struck by the robot. It is clear therefore that the danger area of a robot is not confined to its working volume.

Trapping points can occur within the working volume. It may be possible to be trapped between the links of the manipulator, and at all places where any part of the manipulator and its load approach other fixed items of equipment (which might even be safety guards). In this context then, the danger area is the working volume plus a clearance all round of at least 1 m.

Safety systems

There are three levels of protection against these safety hazards:

- Systems to prevent the operator being in a dangerous situation.
- Systems to protect the operator should this occur.
- Systems to test the second level automatically.

A typical system is illustrated in Figure 12.1.

Guarding, usually of wire mesh or perspex, is the most common means of stopping personnel getting to the danger area but, as mentioned earlier, these structures themselves can present hazards. Guards must be well outside the working volume of the manipulator so that they do not form possible trapping points.

Experience shows that accidents rarely occur with robots under automatic operation: teaching and maintenance present more hazards. For these operations access within the guard enclosure is necessary, so guard gates are interlocked with the robot control system to prevent full-speed automatic operation when a gate is opened. Multirobot installations are guarded in two different ways. Each robot cell may be individually surrounded, with entry to a cell shutting down that particular robot; Figure 12.2(a) illustrates an example of this. Figure 12.2(b) shows a whole line enclosed as one cell—a

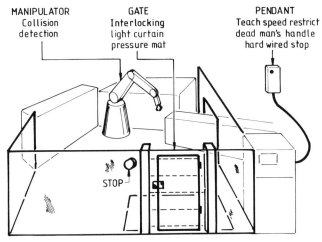

Figure 12.1   Safety systems in a robotic installation.

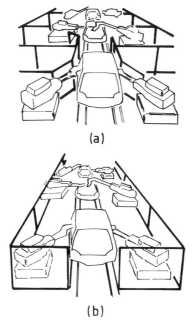

(a)

(b)

Figure 12.2 Robot safety guards: (a) surrounding each robot cell (b) enclosing complete line.

more economical approach in terms of guards, but one that requires all machines to be shut down in order to allow teaching or maintenance work on any one machine in the line.

Teaching by hand-held pendant is potentially more dangerous, since the operator usually needs to be near the end-effector to be able to position it accurately in relation to workpieces and associated machinery. Three aspects of hand-held pendants are considered to be important from a safety point of view:

- 'Dead man's handle': when a pendant is released the manipulator should stop.
- The 'emergency stop' button should be hardwired, and should not rely on software operation.
- Teach speeds should be restricted to safe values.

A second level of safety is necessary to take remedial action should the first level fail or be by-passed. To detect unauthorized entry into the work area, pressure mats and/or light-beam switches are commonly used, the devices being configured to stop the robot and other machinery in the cell when triggered. In certain circumstances it may be appropriate to provide detecting devices on the manipulator itself, so that it will stop if it touches or becomes too close to other equipment. Electrical and pneumatic whiskers, infrared proximity detectors and ultrasonic devices are useful here (see Chapter 8).

It has been suggested (Hunt, 1983) that failsafe design may be achieved by adding a third-level system—one which tests the second level by simulating the conditions to be detected. An example has been quoted of a motor-driven vane used to break the beam to a light switch in order to test the device and its associated circuitry.

12.3   Employment effects

Personal security is affected by health and safety at work: job security has wider ramifications. More than 80 years ago, H. G. Wells said, 'But now that the new conditions physical science is bringing about, not only dispense with man as a source of energy but supply the hope that all routine work may be made automatic, it is becoming conceivable that presently there may be no need for anyone to toil habitually at all...' (Wells, 1905). Wells' vision of Utopia saw beyond the process of mechanization to the time of automation. That time has now come, but its arrival has engendered an industrial climate which is more controversial than utopian. Why is there so much controversy?

A worldwide organization spanning five continents starts to use electronic mail; no controversy arises within the company because no jobs are lost in the company, although there might be related redundancies in some postal organizations. A manufacturing plant introduces a robot to load and unload two CNC lathes; only one of the existing two operators is needed to service the new robot cell, and even that one is not to be fully utilized. Although this may increase the number of jobs in the robot manufacturing industry, the introduction of the robot has directly caused the disappearance of a job in the factory. However, on a global scale, the introduction of a robot is invariably a 'good thing', for efficiency and productivity are improved, and the conflict arises from the opposing interests of management and employees in the immediate future. For management the immediate benefit is increased productivity and/or improved quality; for the employees concerned, the immediate prospect is unemployment. Managers are concerned about the efficient use of manpower, materials, machines and money. Workers are worried about their wages, their hours of work, the length of their holidays, their job security and their working environment. Conflicts are therefore bound to arise.

The immediate and long-term effects of installing robots are recognized by both sides of the labour relations force. The result is that robot manufacturers, industrial engineers, and others who favour the introduction of robots, invariably emphasize the benefits of the exercise to employees—robots taking over dirty or dangerous jobs, job enhancement for those operating the new machinery. Some go so far as to suggest that robots directly create more jobs than they eliminate; in the areas of manufacture, selling, maintenance and operation. On the other hand, the trades unions who represent those employees directly displaced by robots are naturally concerned about job security and often find themselves pushed into a 'we do not resist totally the introduction of new technology as such, but . . .' stance.

Having sketched this scenario of conflict between short-term and long-term gains and losses, we do not propose to comment further on the arguments involved, for that is beyond the scope of this book. The purpose of this section is to identify the quantitative and qualitative effects of the introduction of robots on employment. The effects are usefully divided into direct effects, which are the immediate consequences at the place of introduction of robots, and wider effects, which include the long-term results of the exercise as they affect the user, the short-term increase in demand for robot manufacture and servicing, and the global implications of the process.

Direct effects

Experience suggests that between two and five jobs are lost directly as a result of a robot installation (Williams, 1984; Kalmbach *et al.*, 1982), the losses being manifested as redundancies or, more usually, the displaced operator being transferred within the company and in effect filling a vacancy which would otherwise have brought in another employee. As a counter to this, one or two jobs may be gained in the robot manufacturing industries— manufacture, selling, maintenance and operation. But little of the work may be acquired by the user's company.

From the nature of the jobs being tackled by robots, it is clear that unskilled and semi-skilled workers are likely to be hardest hit by job losses. The demand for these types of workers is steadily declining compared to other occupations.

The bulk of the new work created by a robot installation is likely to be distributed amongst existing employees, and this means that those with craft skills, being more suitable for retraining, are more likely to stay whilst unskilled labour may no longer be required. The new skills needed are in electronics and programming, and it has been shown that 26% of UK companies have a requirement for more people with microelectronic skills, and 11% for those with programming capabilities.

It is argued that these new robot-generated jobs have certain attractive features. They often carry an enhanced status, and the operator also has the more tangible benefit of the retraining which can lead to promotion and increased salary. Against this, the nature of robot installations usually involves more regular cycling of work than was normal with manual operations, and indeed in some applications, e.g. die-casting, this is crucial to the improved quality expected from automatic operation. The regular cycle time imposes a demand upon workers who interface with the robot, and this 'rhythm-obligation' (Kalmbach *et al.*, 1982) can lead to jobs which are more boring because of the monotonous demand of the robot cell.

Wider effects

Considering firstly the global situation, it can be argued that automation in manufacturing creates wealth by reducing unit costs which in turn brings prices down, therefore creating more demand which is met by increased production rates, bringing unit costs even lower, and so round the economic circle. Herein lies the source of the assertion that automation creates more jobs than it eliminates. It has been argued however, that the robotics industry ultimately creates more *activity* than it eliminates in manufacturing, but the number of jobs must inevitably decrease (Obrzut, 1982). Thus the benefits of improving manufacturing industry go to the economy as a whole, rather than specifically within the industry (Fleck, 1983).

There are four aspects of the long-term changes in the pattern of employment which result from the introduction of robots in a plant.

(i) Earlier discussions showed that robot installations can often be financially justified on the basis of existing shift working, which may only be two-shift or even single-shift operation. This means that the future increased demand for products may be met without increases in personnel or most other resources, e.g. buildings. This trend was identified earlier when it was shown that unit costs for robot installation fall as production rates rise, up to the capacity of the plant.

(ii) Experience shows (von Gizycki, 1980) that there is a self-perpetuating element in the automation process. As robots take over jobs previously done manually, more is required from the remaining workers (see above). Their declining ability to fill the automation gaps creates pressure for further automation.

(iii) The robot technician emerges as an important new role in flexible automation. For maintenance work particularly, a comprehensive knowledge of software, electronic, electric and mechanical aspects of a robot installation are required of the maintenance technician. The most pressing reason for this is safety. We have seen that maintenance is a relatively dangerous aspect of a robot since it often involves running the machine with the technician present in the cell. Some companies, e.g. Citroen (Beretti, 1982), now recognize the essentially multidisciplinary role of the robot technician and advocate the requirement of formal qualifications for robot maintenance work.

(iv) Most of the current training in robotics is part of a conversion process for workers moving from the old skills replaced by robots to the robot operation and maintenance sphere. The design of robots itself is a fast-changing phenomenon, and retraining to stay abreast of developments becomes more important. The idea of acquiring a skill sufficient for a lifetime's work has been replaced by a process of constantly updating one's skills as technology progresses. It has been claimed that 'retraining at intervals of ten years will come to be considered the norm for all employees within manufacturing in-dustry'. (Owen, 1982b).

When considering the financial justification of robots, it was noted that all costs are relative. The cost of taking any action must be compared to the cost of taking no action. The same principle, of course, applies to the employment effects of automation. Job losses due to the introduction of robots must be compared to the effect of not introducing robots, and given that competition on an international scale is inexorable it may be the case that maintaining the status quo could have the inevitable consequence of closing a manufacturing plant. The total level of employment is ultimately affected more by the rise and fall of demand, and 'to protect employment through a slow introduction of robotics is clearly missing the point.' (Gerstenfeld, 1982).

12.4 Recapitulation

Making robot installations safer has been presented as a two-stage process: firstly a hazard analysis, and secondly the provision of systems to safeguard against hazards. Impact and trapping are two dangers from the manipulator itself, and the unpredictability of robot movements presents a danger. Safety systems operating at three levels, prevention, protection and test, have been reviewed.

The effects of the introduction of automation have been considered. Long-term benefits accrue to the economy as a whole, and job enhancement is best served by paying attention to training and retraining.

Appendix 1

Lagrangian analysis of a robot with three degrees of freedom

The derivation of the equation presented in Section 5.3 follows. The relevant diagram is repeated here (Figure A1) for convenience.

The method of Lagrange requires the calculation of potential and kinetic energy. We shall consider the kinetic energy first. The coordinates of masses M_2, M_{23}, M_3 and M_{34} are

$$x_2 = 0.5L_2C\theta_2C\theta_1 \qquad x_3 = [L_2C\theta_2 + 0.5L_3C(\theta_2 + \theta_3)]C\theta_1$$
$$y_2 = 0.5L_2C\theta_2S\theta_1 \qquad y_3 = [L_2C\theta_2 + 0.5L_3C(\theta_2 + \theta_3)]S\theta_1$$
$$z_2 = 0.5L_2S\theta_2 \qquad z_3 = L_2S\theta_2 + 0.5L_3S(\theta_2 + \theta_3)$$
$$x_{23} = L_2C\theta_2C\theta_1 \qquad x_{34} = [L_2C\theta_2 + L_3C(\theta_2 + \theta_3)]C\theta_1$$
$$y_{23} = L_2C\theta_2S\theta_1 \qquad y_{34} = [L_2C\theta_2 + L_3C(\theta_2 + \theta_3)]S\theta_1$$
$$z_{23} = L_2S\theta_2 \qquad z_{34} = L_2S\theta_2 + L_3S(\theta_2 + \theta_3)$$

Differentiation of these expressions allows the velocity of each mass to be calculated. For example, for M_2

$$\dot{x}_2 = -0.5L_2[\dot{\theta}_1S\theta_1C\theta_2 + \dot{\theta}_2C\theta_1S\theta_2]$$
$$\dot{y}_2 = 0.5L_2[\dot{\theta}_1C\theta_1C\theta_2 - \dot{\theta}_2S\theta_1S\theta_2]$$
$$\dot{z}_2 = 0.5L_2\dot{\theta}_2C\theta_2$$

The kinetic energy K_2 of M_2 is therefore

$$K_2 = 0.5M_2v_2^2 = 0.5M_2(\dot{x}_2^2 + \dot{y}_2^2 + \dot{z}_2^2)$$

$$K_2 = 0.125M_2L_2^2[\dot{\theta}_1^2C^2\theta_2 + \dot{\theta}_2^2]$$

Following this procedure for each of the masses leads to an expression for the

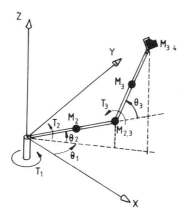

Figure A.1   A robot with three degrees of freedom.

total kinetic energy K:

$$K = 0.125M_2L_2^2[\dot{\theta}_1^2 C^2\theta_2 + \dot{\theta}_2^2] + 0.5M_{23}L_2^2[\dot{\theta}_1^2 C^2\theta_2 + \dot{\theta}_2^2]$$

$$+ 0.5M_3\begin{bmatrix} L_2^2\dot{\theta}_1^2 C^2\theta_2 + L_2^2\dot{\theta}_2^2 + 0.25L_3^2(\dot{\theta}_2 + \dot{\theta}_3)^2 + L_2L_3\dot{\theta}_2(\dot{\theta}_2 + \dot{\theta}_3)C\theta_3 \\ + L_2L_3\dot{\theta}_1^2 C\theta_2 C(\theta_2 + \theta_3) + 0.25L_3^2\dot{\theta}_1^2 C^2(\theta_2 + \theta_3) \end{bmatrix}$$

$$0.5M_{34}\begin{bmatrix} L_2^2\dot{\theta}_1^2 C^2\theta_2 + L_2^2\dot{\theta}_2^2 + L_3^2(\dot{\theta}_2 + \dot{\theta}_3)^2 + 2L_2L_3\dot{\theta}_2(\dot{\theta}_2 + \dot{\theta}_3)C\theta_3 \\ + 2L_2L_3\dot{\theta}_1^2 C\theta_2 C(\theta_2 + \theta_3) + L_3^2\dot{\theta}_1^2 C^2(\theta_2 + \theta_3) \end{bmatrix}$$

The total potential energy P is

$$P = P_2 + P_{23} + P_3 + P_{34}$$

or

$$P = 0.5M_2gL_2S\theta_2 + M_{23}gL_2S\theta_2 + M_3g[L_2S\theta_2 + 0.5L_3S(\theta_2 + \theta_3)]$$
$$+ M_{34}g[L_2S\theta_2 + L_3S(\theta_2 + \theta_3)]$$

The Lagrangian L is defined as

$$L = K - P$$

The torques T_1, T_2 and T_3 can be calculated from

$$T_n = \frac{d}{dt}\frac{\partial L}{\partial \theta_n} - \frac{\partial L}{\partial \theta_n}$$

Appendix 2

Analysis of the parallel-link robot

The parallel-link robot of Section 5.3 was illustrated in Figure 5.7. That figure is repeated here as Figure A2(i) for convenience. Figure A2(ii) shows various projections of the robot's gripper plate in a coordinate frame $x^*y^*z^*$, parallel to the world coordinate frame xyz, and with origin at point 1 on the plate. Point 1 has coordinates $(x_1y_1z_1)$ in the xyz frame.

The coordinates of point 2 are the easiest to determine. Figure A2(ii)(c) gives the x^* and y^* coordinates, and Figure A2(ii)(b) the z^* coordinate of point 2.

$$x_2^* = m \cos \xi \cos \varepsilon$$
$$y_2^* = m \cos \xi \sin \varepsilon$$
$$z_2^* = m \sin \xi$$

where m is the distance between points 1 and 2 on the plate.

Turning to point 4 we can write, using Figure A2(ii)(c)

$$x_4^* = n[\cos \rho \sin \varepsilon + \sin \rho \sin \xi \cos \varepsilon]$$
$$y_4^* = n[\sin \rho \sin \xi \sin \varepsilon - \cos \rho \cos \varepsilon]$$

and from Figure A2(ii)(b)

$$z_4^* = -n \sin \rho \cos \xi$$

where n is the distance from point 1 to point 4, i.e. half the length of a side.

The coordinates of point 3 follow by symmetry:

$$x_3^* = -n[\cos \rho \sin \varepsilon + \sin \rho \sin \xi \cos \varepsilon]$$
$$y_3^* = n[\cos \rho \cos \varepsilon - \sin \rho \sin \xi \sin \varepsilon]$$
$$z_3^* = n \sin \rho \cos \xi$$

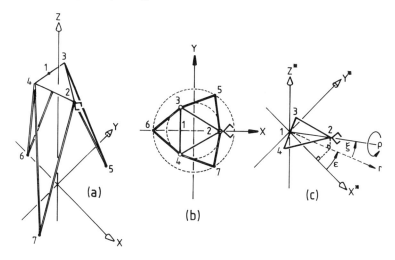

Figure A.2(i)   A parallel topology robot: (a) general arrangement (b) plan view of reference position (c) rotational coordinates.

The same results can be obtained using matrix transformations (Chapter 2). A close examination of Fig. A2(ic) shows that the illustrated configuration can be arrived at by a sequence of rotations about the X^*, Y^* and Z^* axes. Firstly a clockwise rotation ρ about the X^* axis, then an anticlockwise rotation ξ about the Y^* axis and finally a clockwise rotation ε about the Z^* axis. The overall transformation is found

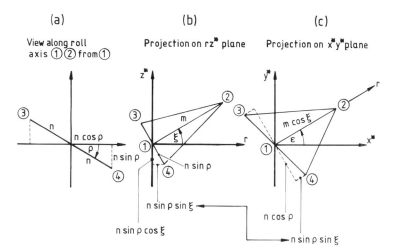

Figure A.2(ii)   Projection of robot's gripper plate in coordinate frame $x^*y^*z^*$: (a) view along roll axis 1 2 from 1 (b) projection on rz^* plane (c) projection on x^*y^* plane.

by multiplying the three rotation transforms.

$$\begin{bmatrix} C\varepsilon & -S\varepsilon & 0 & 0 \\ S\varepsilon & C\varepsilon & 0 & 0 \\ 0 & 0 & 1 & 0 \\ 0 & 0 & 0 & 1 \end{bmatrix} \begin{bmatrix} C\xi & 0 & -S\xi & 0 \\ 0 & 1 & 0 & 0 \\ S\xi & 0 & C\xi & 0 \\ 0 & 0 & 0 & 1 \end{bmatrix} \begin{bmatrix} 1 & 0 & 0 & 0 \\ 0 & C\rho & -S\rho & 0 \\ 0 & S\rho & C\rho & 0 \\ 0 & 0 & 0 & 1 \end{bmatrix}$$

$$= \begin{bmatrix} A & B & C & 0 \\ D & E & F & 0 \\ G & H & I & 0 \\ 0 & 0 & 0 & 1 \end{bmatrix}$$

where $A = C\xi C\varepsilon$ $B = -(C\rho S\varepsilon + S\rho S\xi C\varepsilon)$
       $C = S\rho S\varepsilon - C\rho S\xi C\varepsilon$ $D = C\xi S\varepsilon$
       $E = C\rho C\varepsilon - S\rho S\xi S\varepsilon$ $F = -(S\rho C\varepsilon + C\rho S\xi S\varepsilon)$
       $G = S\xi$ $H = S\rho C\xi$ $I = C\rho C\xi$

References

Abele, E. (1981) Adaptive controls for fettling of castings with an industrial robot, *Proc. 1st Internat. Conf. Robot Vision and Sensory Controls.*

Airey, J. and Young, C. (1983) Economic justification—counting the strategic benefits, *Proc. 2nd Internat. Conf. Flexible Manufacturing Systems.*

Akeel, H. A. (1982) Expanding the capabilities of spray painting robots, *Robotics Today,* April, pp. 50–53.

Albus, J. S. (1981) *Brains, Behaviour and Robotics,* McGraw-Hill.

Albus, J. S., McLean, C. R., Barbera, A. J. and Fitzgerald, M. L. (1983) Hierarchical control for robots in an automated factory, *Proc. 13th. Internat. Symp. Industrial Robots.*

Aleksander, I. (1983) *Artificial Vision for Robots,* Kogan Page.

Aleksander, I. and Burnett, P. (1983) *Reinventing Man,* Kogan Page.

Allan, R. (1983) Tactile sensing, 3-D vision, and more precise arm movements herald the hardware trends in industrial robots, *Electronic Design,* 12 May, pp. 99–112.

Ambler, A. P., Popplestone, R. J. and Kempf, K. G. (1982) An experiment in the offline programming of robots, *Proc. 12th Internat. Symp. Industrial Robots.*

Arzebaecher, R. C. (1960) Servomechanisms with force feedback, ANL 6157.

Asada, H., Kanade, T. and Takeyama, I. (1983) A direct-drive manipulator: Development of a high-speed manipulator, in *Developments in Robotics,* IFS Publications, (1983).

Automatix (1982) RAIL Software Reference Manual (ROBOVISION and CYBERVISION), Rev. 3.0, MN-RB-07, Automatix Inc.

Baker, J. M. (1981) How to achieve recognition, *Speech Technology,* Fall, pp. 30–43.

Barrow, H. G. (1980) Artificial Intelligence, Infotech State of the Art Report Series 8, No. 6, Factory Automation, Infotech Ltd.

Baruh, I. S., Konstantinov, M. M., Delov, J. P. and Patarinski, S. P. (1980) Microprocessor based algorithms for control of industrial robots, *Proc. Control Problems and Devices in Manufacturing Technology,* IFAC.

Bauzil, G., Briot, M. and Ribes, R. (1981) A navigation sub-system using ultrasonic sensors for the mobile robot Hilare, *Proc. 1st Internat. Conf. Robot Vision and Sensory Control.*

Bejczy, A. K. (1974) Robot arm dynamics and control, Technical Memo 33-669, Jet Propulsion Laboratory.

Bejczy, A. K. (1980) Sensors, controls and man-machine interface for advanced teleoperation, *Science* **208,** No. 4450, pp. 1327–1335.

Belforte, G., D'Alfio, N., Quagliotti, F. and Romiti, A., (1981) Identification through air jet sensing, *Proc. 1st Internat. Conf. Robot Vision and Sensory Controls.*

Beretti, A. (1982) Robotics: a new career opportunity for technicians in industry, *Proc. 12th Internat. Symp. Industrial Robots.*

Bessonov, A. P. and Umnov, N. W. (1976) Choice of geometric parameters of walking machines, *Proc. 2nd Symp. Theory and Practice of Robots and Manipulators.*

Bessonov, A. P. and Unmov, N. V. (1983) The stabilization of the position of the body of walking machines, *Mech. Mach. Theory* **18**, No. 4, pp. 261–265.

Boucharlat, G., Chabbal, J. and Chautemps, J., (1984) 256×256 pixel CCD solid state image sensor, *Proc. 4th Internat. Conf. Robot Vision and Sensory Controls.*

BRA (1984) 1984/85 UK Robots Industry Directory, British Robot Association.

BRA (1985) *Robot Facts 1984*, British Robot Association.

Braggins, D. (1984) Giving robots sense. *Automation*, November, pp. 32–34.

Bryce, A. L. G. (1983) Is there such a thing as low-cost FMS? *Proc. 2nd Internat. Conf. Flexible Manufacturing Systems.*

Burnett, J. R. (1957) Force reflecting servos add feel to remote controls, *Control Engng* **4**, 82–87.

Busby, R. F. (1976) Manned Submersibles, Office of the Oceanographer of the Navy (USA).

C & I (1986) Various articles on programmable controllers, *Control and Instrumentation*, Vol. 18, No. 1, January, 39–71.

CADCAM (1982) *CADCAM: Your Questions Answered*, The CADCAM Association.

CADCAM (1983) *A Guide to CADCAM*, Institution of Production Engineers/The Numerical Engineering Society.

Campbell, N. A., Reid, I. M. and McClean, J. H. (1984) Short's robot ultrasonic scanning system, in *UK Robotics Research 1984*, MEP.

Carlisle, B., Roth, S., Gleason, J. and McGhie, D. (1981) The PUMA TM/VS-100 robot vision system, *Proc. 1st Internat. Conf. Robot Vision and Sensory Controls.*

Carter, C. F. (1972) Trends in machine tool development and application, *Proc. 2nd Internat. Conf. Product Development and Manufacturing Technology.*

Charles, J. and Vertut, J. (1977) Cable controlled deep submergence tele-operator system, *Mech. Mach. Theory* **12**, pp. 481–492.

Chechinski, S. S. and Agrawal, A. K. (1983) Magnetoelastic tactile sensor, *Proc. 3rd. Internat. Conf. Robot Vision and Sensory Controls.*

Chen, Fan Yu (1982) Gripping mechanisms for industrial robots, *Mech. Mach. Theory* **17**, No. 5, pp. 299–311.

Childress, D. S. (1973) Powered limb prostheses—their clinical significance, *IEEE Trans. Biomed. Engng* **20**, No. 3, pp. 200–207.

Cincinnati (1980a) Operating Manual—T3 Industrial Robot, Version 3.0, Publication No. 1-IR-79149, Cincinnati Milacron.

Cincinnati (1980b) Basic Documentation Acramatic Robot Control (RC) V3 Hardware, Publication No. 7-000-0393AA, Vols 1–3, Cincinnati Milacron.

Clowes, M. B. (1971) On seeing things, *Artificial Intelligence* **1.**

Coiffet, P. (1983a) *Robot Technology Vol. 1: Modelling and Control*, Kogan Page.

Coiffet, P. (1983b) *Interaction with the Environment*, Kogan Page.

Colson, J. C. and Perreira, N. D. (1983) Kinematic arrangements used in industrial robots, *Proc. 13th Internat. Symp. Industrial Robots.*

Cook, G. E. (1983) Position sensing with an electric arc, *Proc. 13th Internat. Symp. Industrial Robots.*

Corell, R. W. and Wijnschenk, M. J. (1964) Design and development of the Case Research Arm-Aid, Case Institute of Technology Report EDC 4-64-4.

Crossley, T. R. and Lo, E. K. K. (1981) Key lecture on robotics, 2nd Special Convention on Information Technologies, Productivity and Employment.

Cugy (1984) *Industrial Robot Specifications*, Kogan Page.

Dapiran, A. and Manieri, M. (1983) The cost of flexibility in the FMS, *Proc. 2nd Internat. Conf. Flexible Manufacturing Systems.*

Dario, P., Domenici, C., Bardelli, R., De Rossi, D. and Pinotti, P. C. (1983) Piezoelectric polymers: new sensor materials for robotic applications, *Proc. 13th Internat. Symp. Industrial Robots*.

Datta, I. and Kuo, C. (1978) The role of manipulators for underwater activities offshore, *Proc. 3rd Symp. Theory and Practice of Robots and Manipulators*.

Davies, E. R. (1984a) Design of cost-effective systems for the inspection of certain food products during manufacture, *Proc. 4th Internat. Conf. Robot Vision and Sensory Controls*.

Davies, E. R. (1984b) A glance at image analysis, *Chartered Mech Engnr*, Vol. 31, No. 12, pp. 32–35.

Denavit, J. and Hartenberg, R. S. (1955) A kinematic notation for lower pair mechanisms based on matrices, *ASME J. Appl. Math.*, Vol. 77, pp. 215–221.

Dillman, R. (1982) A sensor controlled gripper with tactile and non-tactile sensor environment, *Proc. 2nd Internat. Conf. Robot Vision and Sensory Control*.

Doebelin, E. O. (1983) *Measurement Systems: Application and Design*, McGraw-Hill.

Donato, G. and Camera, A. (1981) *A High Level Programming Language for a New Multi-arm Robot Assembly*, DEA.

During, J. and van Miltenburg, T. C. M. (1967) A emg operated control system for a prosthesis, *Med. Biol Engng* **5**, no. 6, pp. 597–601.

ECC (1972) *DC Motors: Speed Controls, Servo Systems*, Electrocraft Corporation, Pergamon.

Edling, G. and Porsander, T. (1984) Adaptive control of torch position and welding parameters in robotic arc welding—examples and practical use, *Proc. 4th Internat. Conf. Robot Vision and Sensory Controls*.

Edson, D. (1984) Bin picking robots punch in, *High Technology*, June, pp. 57–60.

Engleberger, J. F. (1980) *Robotics in Practice*, Kogan Page.

Erdelyi, F., Nemes, L. and Orban, P. (1980) Path calculation and sampled data control systems for multi-axis machines, *Proc. Control Problems and Devices in Manufacturing Industry Conf.*, IFAC.

Espiau, B. and Catros, J. Y. (1980) Use of optical reflectance sensors in robotics applications, *IEEE Trans. Syst. Man Cybernet*, Vol. SMC-10, No. 12, December, pp. 901–912.

Evard, F., Farreny, H. and Prade, H. (1982) A pragmatic interpreter of a task-oriented subset of natural language for robotic purposes, *Proc. 12th Internat. Symp. Industrial Robots*.

Fleck, J. (1983) The adoption of robots, *Proc. 13th Internat. Symp. Industrial Robots*.

Fletcher, M. J. and Leonard, F. (1955) The principles of artificial hand design, *Artificial Limbs* **2**.

Fountain, T. (1983) Image processing by parallel computer, *Automation*, September, pp. 8–15.

Fryer, R. J. (1984) Archie—an experimental 3D vision system, *Proc. 4th Internat. Conf. Robot Vision and Sensory Controls*.

Fujiwara, K., Kawashima, Y., Kato, H. and Watanabe, M. (1977) Development of guideless robot vehicle, *Proc. 7th Internat. Symp. Industrial Robots*.

Gaillet, A. and Reboulet, C. (1983) An isostatic six component force and torque sensor, *Proc. 13th Internat. Symp. Industrial Robots*.

Gandy, T. G. (1983) A simple robot system for loading/unloading internal grinders, *Proc. 13th Internat. Symp. Industrial Robots*.

GEC (1968) Hardiman prototype project, Report S–68–1060, General Electric Co.

Gerstenfeld, A. and Berger, P. (1982) A model for economic and social evaluation of industrial robots, *Proc. 12th Internat. Symp. Industrial Robots*.

Goertz, R. C. (1952) Fundamentals of general purpose remote manipulators, *Nucleonics* **10**.

Goertz, R. C. (1964) Some work on manipulator systems at ANL: past, Present and a look at the future, *Proc. 1964 Seminars Remotely Operated Special Equipment* Vol. 1, AEC Conf. 640508.

Goertz, R. C., Burnett, J. R. and Bevilacqua, F. (1953) Servos for remote manipulation, AEC ANL–5022.

Gough, V. E. and Whitehall, S. G. (1962) Universal tyre test machine, *Proc. 9th Internat. Conf. FISITS,* Institute of Mechanical Engineers.

Gruver, W. A., Soroka, B. I., Craig, J. J. and Turner, T. L. (1983) Evaluation of commercially available robot programming languages, *Proc. 13th Internat. Symp. Industrial Robots.*

Guitte, J., Kwee, H. H., Quetin, N. and Yelon, J. (1979) The Spartacus telethesis manipulator control studies, *Bull. Prosth. Res.,* BPR 10-32, Fall, pp. 69–105.

Gunn, T. G. (1982) The mechanization of design and manufacturing, *Scientific American,* September.

Gustavson, R. E. (1983) Choosing manufacturing systems based on unit cost, *Proc. 13th Internat. Symp. Industrial Robots.*

Guzman, A. (1968) Computer recognition of three dimensional objects in a scene, MIT Report MAC-TR-59.

Hareland, A. (1983) No novice in surface coating, in *Decade of Robotics,* IFS (1983).

Harmon, L. D. (1982) Automated tactile sensing *Int. J. Robotics Res.* **1,** No. 2, Summer, pp. 3–32.

Harris, D. M. J. and Irvine, D. A. (1984) Robot sensory systems, *Proc. 1st Conf. Irish Manufacturing Committee: Manufacturing Technology: Research and Development,* Parsons Press.

Hartley, J. (1983) *Robots at Work,* IFS Publications.

Hartley, J. (1984) Semi-direct drive to the fore. *The Industrial Robot,* September, pp. 158–161.

Hasegawa, Y. (1979) New developments in industrial robots. *Int. J. Production Res.* August.

Hasegawa, Y. and Sugimoto, N. (1982) Industrial safety and robots, *Proc. 12th Internat. Symp. Industrial Robots.*

Haynes, L. S., Barbera, A. J., Albus, J. S., Fitzgerald, M. L. and McCain, H. G. (1984) An application example of the NBS robot control system, *Robotics and Computer-Integrated Manufacturing* **1,** No. 1, pp. 81–95.

Heathkit (1980) *Industrial Robots and Electronics,* Heathkit Educational Systems.

Hirose, S. and Umetani, Y. (1977) The development of soft gripper for the versatile robot hand, *Proc. 7th Internat. Symp. Industrial Robots.*

Hirose, S. and Umetani, Y. (1978) Some considerations on a feasible walking mechanism as a terrain vehicle, *Proc. 3rd Symp. Theory and Practice of Robots and Manipulators.*

Hirose, S. and Umetani, Y. (1981) A cartesian co-ordinates manipulator with articulated structure, *Proc. 11th Internat. Conf. Industrial Robots.*

Hnatek, E. R. (1976) *A User's Handbook of D/A and A/D Converters,* Wiley.

Hohn, R. E. (1978) Computed path control for an industrial robot, *Proc. 8th Internat. Symp. Industrial Robots.*

Holland, S. W., Rossol, L. and Ward, M. R. (1979) Consight-1: a vision-controlled robot system for transferring parts from belt conveyors, in *Computer Vision and Sensor Based Robots,* Plenum Press.

Horner, G. R. and Lacey, R. J. (1983) High performance brushless PM motors for robotics and actuator applications, *Proc. 1st European Conf. on Electrical Drives/Motors/ Controls S2,* PLL Conference Publications No. 19.

Hotta, H., Ohtsuka, K. and Natori, K. (1977) Underwater bilateral servo manipulator, *Proc. 7th Internat. Symp. Industrial Robots.*

Houldcroft, P. T. (1979) *Welding Process Technology,* Cambridge University Press.

Hrones, J. A. and Nelson, G. L. (1951) *Analysis of the Four-Bar Linkage,* MIT Press.

Huffman, D. A. (1971) Impossible objects as nonsense sentences, in *Machine Intelligence* **6,** Edinburgh University Press.

Hunt, K. H. (1978) *Kinematic Geometry of Mechanisms,* Clarendon Press.

Hunt, K. H. (1982) Geometry of robotic devices, *Inst. Engrs Aust., Mech. Engng Trans.,* pp. 213–220.

Hunt, V. D. (1983) *Industrial Robotics Handbook,* Industrial Press.

Hutchinson, A. C. (1967) Machines can walk, *Chartered Mech. Engnr.,* November, pp. 480–484.

IBM (1981) *AML Reference Manual,* 2nd edn, IBM.

IFS (1983) *Decade of Robotics,* IFS Publications.

Jacobsen, S. C., Knutti, D. F., Johnson, R. T. and Sears, H. H. (1982) Development of the Utah artificial arm, *IEEE Trans. Biomed. Engng* **4,** Vol. BME-29, No. 4., April, pp. 249–269.

Johnsen, E. G. and Corliss, W. R. (1969) *Human Factors Applications in Tele-operator Design and Information,* Wiley-Interscience.

Johnsson, U., Almström, C., Körner, L., Herberts, P. and Kadefors, R. (1981) A microprocessor based system for control of multifunctional myoelectric hands, *Proc. 7th Internat. Conf. External Control of Human Extremities.*

Jones, B. M. and Saraga, P. (1981) The application of parallel projections to three dimensional object location in industrial assembly, *Pattern Recognition* **14.**

Kafrissen, E. and Stephans, M. (1984) *Industrial Robots and Robotics,* Reston/Prentice Hall.

Kalmbach, P., Kasiske, R., Manske, F., Mickler, O., Pelull, W. and Wobbe-Ohlenburg, W. (1982) Robots effect on production, work and employment, *The Industrial Robot,* March, pp. 42–45.

Kasai, M., Takeyasu, K., Uno, M. and Muraoka, K. (1981) Trainable assembly system with an active sensory table possessing six axes, *Proc. 11th Internat. Symp. Industrial Robots.*

Kato, I. (1977) *Mechanical Hands,* Tokyo.

Kauffman, H. (1983) ASEA's secondary current development transmission system for spot welding systems, in *Developments in Robotics,* IFS Publications.

Kay, J. (1983) Proximity switches, *Electrical Equipment,*

Kelley, C. R. (1968) *Manual and Automatic Control,* Wiley.

Konstantinov, M. S. (1975) Mechanical grippers, *Proc. 5th. Internat. Symp. Industrial Robots.*

Kremers, J., Blahnik, C., Brain, A., Cain, R., DeCurtins, J., Meseguer, J. and Peppers, N. (1983) Development of a machine-vision-based robotic arc welding system, *Proc. 13th Internat Symp. Industrial Robots.*

Kretch, S. J. (1982) Robotic animation, *Mech. Engng,* August, pp. 32–35.

Kuwahara, Y. (1982) The Japanese way of robot life, *Employment Gazette,* August, pp. 346–350.

Lanton, S. and Sakasai, B. (1983) Low cost absolute precision data for robots, *Proc. 13th Internat. Symp. Industrial Robots.*

Larcombe, M. H. E. (1981) Carbon fibre tactile sensors, *Proc. 1st Internat. Conf. Robot Vision and Sensory Control.*

Lee, C. S. G. (1982) Robot arm kinematics, dynamics and control, *Computer,* December.

Lee, C. S. G. and Ziegler, M. (1983) A geometric approach in solving the inverse kinematics of PUMA robots, *Proc. 13th Symp. Industrial Roots.*

Lefer, H. (1970) Electrohydraulic servos control walking machines, *Hydraulics and Pneumatics,* August, pp. 63–65.

Levas, A. and Selfridge, M. (1983) Voice communications with robots, *Proc. 13th Internat Symp. Industrial Robots.*

Lewis, A., Nagpal, B. I. C. and Watts, P. L. (1983) Investment analysis for robotic applications, *Proc. 13th Internat. Symp. Industrial Robots.*

Lewis, R. A. and Johnson, A. R. (1977) A scanning laser range finder for a robot vehicle, *Proc. 5th Internat. Joint Conf. Artificial Intelligence.*

Lian, D., Peterson, S. and Donath, M. (1983) A three-fingered, articulated, robotic hand, *Proc. 13th Internat. Symp. Industrial Robots.*

Lieberman, L. I. and Wesley, M. A. (1977) AUTOPASS: An automatic programming system for computer controlled mechanical assembly, *Int. J. Res. Devel.* **21,** Pt. 4, pp. 321–333.

Lhote, F., Kauffmann, J. M., Andre, P. and Taillard, J. P. (1984) *Robot Components and Systems,* Kogan Page.

Luh, J. Y. S., Walker, M. W. and Paul, R. P. C. (1979) Newton-Euler formulation of manipulator dynamics for computer control, *Information and Control Problems in Manufacturing Technology Conference,* IFAC.

Luh, J. Y. S., Walker, M. W. and Paul, R. P. C. (1980a) On-line computational scheme for mechanical manipulators, *J. Dynamic Sys. Meas. Control* **102**, No. 2, June, pp. 69–76.

Luh, J. Y. S., Walker, M. W. and Paul, R. P. C. (1980b) Resolved acceleration control of mechanical manipulators, *IEEE Trans. Autom. Control*, Vol. 25, No. 3, June, pp. 468–474.

Lundstrom, G. (1976) A new method of designing grippers, *Proc. 6th Internat. Symp. Industrial Robots*.

Lundstrom, G., Glemme, B. and Rooks, B. W. (1977) *Industrial Robots—Gripper Review*, IFS Publications.

Luo, R. C., Suresh, S. and Grande, D. (1983) Sensors for cleaning castings with robot and plasma arc, *Proc. 3rd Internat. Conf. Robot Vision and Sensory Controls*.

Makino, H. and Furuya, Y. (1982) SCARA robot and its family, *Proc. 3rd Conf. on Assembly Automation*.

Margrain, P. (1983) Servo-actuators and robotics, in *Developments in Robotics*, IFS Publications.

Martins, G., Nordström, R. and Svensson, M. (1983) Machining cells put robots in their right place, in *Decade of Robotics*, IFS (1983).

Masuda, R. and Hasegawa, K. (1981) Total sensory system for robot control and its design approach, *Proc. 12th Internat. Symp. Industrial Robots*.

McCloy, D. (1984) *Technology Made Simple*, Heinemann.

McCloy, D. and Martin, H. R. (1980) *The Control of Fluid Power*, Ellis Horwood.

McCormack, W. and Godding, E. G. (1983) The creeping technology of grinding and fettling, in *Decade of Robotics*, IFS (1983).

McGhee, R. B. (1968) Some finite state aspects of legged locomotion, *Math. Biosci.* **2**, pp. 67–84.

McGhee, R. B., Chao, C. S., Jaswa, V. C. and Orin, D. E. (1978) Real time computer control of a hexapod vehicle, *Proc. 3rd Internat. Symp. Theory and Practice of Robots and Manipulators*.

McGhee, R. B. and Frank, A. A. (1968) On the stability properties of quadruped creeping gaits, *Math. Biosci.* **3**, pp. 331–351.

McGhee, R. B. and Iswandhi, G. I. (1979) Adaptive locomotion of a multilegged robot over rough terrain, *IEEE Trans. Syst. Man Cybernet.* **9**, Vol. SMC-9, No. 4, April, pp. 176–182.

McGhee, R. B. and Jain, A. K. (1972) Some properties of regularly realizable gait matrices, *Math. Biosci.* **13**, pp. 179–193.

McGovern, D. E. (1977) An investigation of supervisory control of remote manipulation, *Mech. Mach. Theory* **12**, pp. 3–9.

Megahed, S. and Renaud, M. (1982) Minimization of the computation time necessary for the dynamic control of robot manipulators, *Proc. 12th Internat. Symp. Industrial Robots*.

Menadier, C. *et al.* (1967) The Fotonic Sensor, *Instrument and Control Systems*, Mechanical Technology Inc., Latham, NY.

Milenkovic, V. (1979) Computer synthesis of continuous path robot motion, *Proc. 5th World Congr. Theory of Machines and Mechanisms*.

Milenkovic, V. and Huang, B. (1983) Kinematics of major robot linkages, *Proc. 13th Conf. Industrial Robots*.

Mizutame, M., Sugiyama, S., Takaishi, K., Tomioka, K. and Sakurai, T. (1984) Development of robot-press production system, *Proc. 24th Internat. Machine Tool Design and Research Conf.*

Money, S. A. (1982) *Microprocessor Data Book*, Granada.

Morgan, C. G., Bromley, J. S. E., Davey, P. G. and Vidler, A. R. (1983) Visual guidance techniques for robot arc welding, *Proc. 3rd Internat. Conf. Robot Vision and Sensory Controls*.

Morganite (1961) *Carbon Brushes and Electrical Machines*, Morganite Carbon Ltd.

Mosher, R. S. (1967) Handyman to Hardiman, SAE Paper 670088.

Mosher, R. S. and Wendel, B. (1960) Force reflecting electrohydraulic servomanipulator, *Electrotechnology* **66**, 138–141.

Mott, D. H., Lee, M. H. and Nicholls, H. R. (1984) An experimental very high resolution tactile sensor array, *Proc. 4th Internat. Conf. Robot Vision and Sensory Controls.*

MTTA (1982) *Safeguarding Industrial Robots Part 1—Basic Principles,* The Machine Tool Trades Association.

Mujtaba, S. M. (1980) Current status of the AL manipulator programming system, *Proc. 10th Internat. Symp. Industrial Robots.*

Mujtaba, S. M. and Goldman, R. (1979) The AL User's Manual, STAN-CS-79-719, Stanford University.

Nally, G. M. (1983) Robotic arc welding: at what state is the art? *Robotics Today,*

Napier, J. (1967) The antiquity of human walking, *Sci. Am.* **216,** No. 4, April, pp. 56–66.

Nathan, C. A. and Flatau, C. R. (1978) Manipulators for large scale construction in outer space, *Proc. 26th Conf. Remote Systems Technology.*

Neubauer, G. (1982) Pneumatic grippers for a vise-like grip or a gentle squeeze, *Machine Design,* November 25, pp. 69–71.

Nevatia, R. (1982) *Machine Perception,* Prentice Hall.

Nevins, J. *et al.* (1974) A scientific approach to the design of computer controlled manipulators, Report R 837, Charles Stark Draper Laboratory, MIT.

Nevins, J. L. and Whitney, D. E. (1977) Information and control issues of adaptable programmable assembly systems for manufacturing and tele-operator applications, *Mech. Mach. Theory* **12,** pp. 27–43.

Nimrod, N., Margalith, A. and Mergler, H. W. (1982) A laser-based scanning range finder for robotic applications, *Proc. 2nd Internat. Conf. Robot Vision and Sensory Controls.*

Nitzan, D., Brain, A. E. and Duda, R. O. (1977) The measurement and use of registered reflectance and range data in scene analysis, *Proc. IEEE* **65,** No. 2, February, pp. 206–220.

Northcott, J. and Rogers, P. (1984) *Microelectronics in British Industry: The Pattern of Change,* No. 625, Policy Studies Institute.

Obrzut, J. J. (1982) Robotics extends a helping hand, *Iron Age,* 19 March, pp. 60–83.

Ogo, K., Ganse, A. and Kato, I. (1978) Quasi dynamic walking of biped walking machine aimed at completion of steady walking, *Proc. 3rd Symp. Theory and Practice of Robots and Manipulators.*

Okada, T. and Tsuchiya, S. (1977) On a versatile finger system, *Proc. 7th Internat. Symp. Industrial Robots.*

Onega, G. T. and Clingman, J. H. (1972) Free flying tele-operator requirements and conceptual design, *Proc. 1st Nat. Conf. Remotely Manned Systems.*

Owen, A. E. (1982a) Automated assembly can equate with short payback periods, *Proc. 3rd Internat. Conf. Assembly Automation.*

Owen, A. E. (1982b) *Chips in Industry,* Special Report No. 135, The Economist Intelligence Unit.

PERA (1982) Vision Systems, PERA Report 366, Production Engineering Research Association.

Park, W. T. (1981) *The SRI Robot Programming System (RPS)—An Executive Summary,* SRI International.

Patla, A. E., Hudgins, B. S., Parker, P. A. and Scott, R. N. (1982) Myoelectric signal as a quantitative measure of muscle mechanical output, *Med. Biol. Engng Comput.* **20,** May, pp. 319–328.

Patzelt, W. (1982) A robot position control algorithm for the grip onto an accelerated conveyor or belt, *Proc. 12th Internat. Symp. Industrial Robots.*

Paul, R. P. (1982) *Robot Manipulators: Mathematics, Programming and Control,* MIT Press.

Paul, R. P., Shimano, B. and Mayer, G. E. (1981) Kinematic control equations for simple manipulators, *IEEE Trans. Syst. Man Cybernet* **11,** No. 6, June.

Pavone, V. J. (1983) User friendly welding robots, *Proc. 13th Internat. Symp. Industrial Robots.*

Peizer, E. *et al.* (1969) The Otto Bock hand, *Bull. Prosth. Res.*

Percival, N. (1984) Safety aspects of industrial robots, *Metal Construction,* April, pp. 201–203.

Pham, D. T. (1983) A low cost industrial robot, in *Developments in Robotics*, IFS Publications.

Polaroid (1980) Polaroid Ultrasonic Ranging Experimenter's Kit, Polaroid Corporation.

Popplestone, R. J. *et al.* (1978) RAPT: A language for describing assemblies, in *Coll. On Robotics*, National Engineering Laboratory.

Potkonjak, V., Vukobratovic, M. and Hristic, D. (1983) Interactive procedure for computer-aided design of industrial robot mechanisms, *Proc. 13th Internat. Symp. Industrial Robots*.

Presern, S., Spegel, M. and Dzinek, I. (1981) Tactile sensing with sensory feedback control for industrial arc welding robots, *Proc. 1st Internat. Conf. Robot Vision and Sensory Controls*.

Pugh, A. (1983) Second generation robotics and robot vision, Chapter 1 of Robotic Technology, Short Run Press.

Purbrick, J. A. (1981) A force transducer employing conductive silicone rubber, *Proc 1st Internat. Conf. Robot. Vision and Sensory Controls*.

Raibert, M. H. and Sutherland, I. E. (1983) Machines that walk, *Sci. Am.*, January, pp. 32–41.

Raibert, M. H. and Tanner, J. E. (1982) A VLSI tactile array sensor, *Proc. 12th Internat. Symp. Industrial Robots*.

Ranky, P. G. (1983) *The Design and Operation of FMS*, IFS Publications.

Ray, J. (1983) Let the application choose the drive, in *Decade of Robotics*, IFS (1983).

Rebman, J. and Morris, K. A. (1983) A tactile sensor with electrooptical transduction, *Proc. 3rd Internat. Conf. Robot Vision and Sensory Controls*.

Reed, C. K. (1979) Two hands are better than one, *Proc. 9th Internat. Symp. Industrial Robots*.

Roberts, L. G. (1963) Machine perception of three-dimensional solids, in *Optical and Electro-optical Information Processing*, MIT Press.

Robertson, B. E. and Walkden, A. J. (1983) Tactile sensor system for robotics, *Proc. 3rd Internat Symp. Robot Vision and Sensory Controls*.

Rosen, C. A. (1979) Machine vision and robotics: industrial requirements, in *Computer Vision and Sensor Based Robots*, Plenum Press.

Ross, M. H. (1981) Automated manufacturing—why is it taking so long? *Long Range Planning* **14**, No. 3, pp. 28–35.

Rovetta, A., Vicentini, P. and Franchetti, I. (1981) On development and realization of a multipurpose grasping system, *Proc 11th Internat. Conf. Industrial Robots*.

Ruder, H. (1982) A modern assembly robot, *Proc. 3rd Internat. Conf. Assembly Automation*.

Schmalenbach, E., Küppers, H.-J. and Roesler, H. (1978) Manipulators for severely handicapped: control philosophy and applications, *Proc. 3rd Symp. Theory and Practice of Robots and Manipulators*.

Schmidl, H. (1977) The I.N.A.I.L. experience fitting upper limb dysmelia patients with myoelectric control, *Bull. Prosth. Res.*, Spring, pp. 17–42.

Scott, P. B. and Husband, T. M. (1983) Robotic assembly: Design,, analysis and economic evaluation, *Proc. 13th Internat. Symp. Industrial Robots*.

Shirai, Y. (1972) Recognition of polyhedra with a range finder, *Pattern Recognition* **4.**

Sigma (1980) *Sigma Programming Handbook*, Olivetti SpA.

Skinner, F. (1975) Designing a multiple prehension manipulator, *Mech. Engng*, September, pp. 30–37.

Stackhouse, T. (1979) A new concept in robot wrist flexibility, *Proc. 9th Internat. Symp. Industrial Robots*.

Stauffer, R. N. (1983a) Progress in tactile sensor development, *Robotics Today*, June, pp. 43–49.

Stauffer, R. N. (1983b) Robots provide efficiency and quality in new production welding lines at Jeep, *Manufacturing Engng*, October, pp. 57–60.

Stewart, D. (1965) A platform with six degrees of freedom, *Proc. Inst. Mech. Engrs* **180**, Pt. 1, No. 15, pp. 371–386.

Sugimoto, N. and Kawaguchi, K. (1983) Fault tree analysis of hazards created by robots, *Proc. 13th Internat. Symp. Industrial Robots.*

Swain, I. D. (1982) Adaptive Control of an Arm Prosthesis, Ph.D. Thesis, University of Southampton.

Swain, I. D. and Nightingale, J. M. (1981) Microprocessor control of a multifunctional arm and hand prosthesis, *Proc. 7th Internat. Conf. External Control of Human Extremities.*

Sword, A. J. and Park, W. T. (1977) Location and acquisition of objects in unpredictable locations, *Mech. Mach. Theory* **12**, pp. 123–132.

Taguchi, K., Ikeda, K. and Matsumoto, S. (1976) Four-legged walking machine, *Proc. 2nd Symp. Theory and Practice of Robots and Manipulators.*

Takano, M. and Odawara, G. (1981) Development of new type of mobile robot TO-ROVER, *Proc. 11th Internat. Symp. Industrial Robots.*

Takata, S. and Kishimoto, Y. (1977) Fire fighting robot, *Proc. 7th Internat. Symp. Industrial Robots.*

Tanil, K., Komoriya, K., Kaneko, M., Tachi, S. and Fujikawa, A. (1984) A high resolution tactile sensor, *Proc. 4th Internat. Conf. Robot Vision and Sensory Controls.*

Tarvin, R. L. (1980) Considerations for off-line programming a heavy duty industrial robot, *Proc. 10th Internat. Symp. Industrial Robots.*

Taylor, M. J. (1978) Large scale manipulator for space shuttle payload handling, *Proc. 26th Conf. Remote Systems Technology.*

Thring, M. W. (1983) *Robots and Telechirs,* Ellis Horwood.

Toyama, S. and Takano, M. (1981) Study on speed-up of robot motion, *Proc. 11th Internat. Symp. Industrial Robots.*

Trounov, A. N. (1984) Applications of sensory models for adaptive robots, *Proc. 4th Internat. Conf. Robot Vision and Sensory Controls.*

Turner, T. L., Craig, J. J. and Gruver, W. A. (1984) A microprocessor architecture for advanced robot control, *Proc. 14th Internat. Symp. Industrial Robots.*

VAL (undated) *User's Guide to VAL,* Version 12, No. 398-H2A, Unimation.

Vicentine, P. (1983) Robotics changes the production system and social aspects of work, *Ital. Mach. Equip.,* March.

Volz, R. A., Mudge, T. N. and Gal, D. A. (1983) Using ADA as a robot system programming language, *Proc. 13th Internat. Symp. Industrial Robots.*

von Gizycki, R. (1980) Social conditions for and consequences of the use of industrial robots in five factories, *Proc. 10th Internat. Symp. Industrial Robots.*

Vranish, J. M. (1984) Magnetoresistive skin for robots, *Proc. 4th Internat. Conf. Robot Vision and Sensory Controls.*

Vukobratović, M. (1973) How to control artificial anthropomorphic systems, *IEEE Trans. Syst. Man Cybernet.* Vol. SMC-3, No. 5, September, pp. 497–507.

Waltz, D. L. (1975) Understanding line drawings of scenes with shadows, in *The Psychology of Computer Vision,* McGraw-Hill.

Wang, S. S. M. and Will, P. M. (1978) Sensors for computer controlled mechanical assembly, *The Industrial Robot,* March, pp. 9–18.

Warnecke, H.-J. and Schraft, R. D. (1979) *Industrieroboter,* Krausskopt-Verlag, Mainz.

Warnecke, H.-J. and Schraft, R. D. (1982) *Industrial Robots Application Experience,* IFS Publications.

Wells, H. G. (1905) *A Modern Utopia,* Chapman and Hall.

Whitney, D. E. and Nevins, J. L. (1979) What is the remote center compliance and what can it do? *Proc. 9th Internat. Symp. Industrial Robots.*

Wickman (undated) *Wickman Programmable Logic Systems,* Wickman Automation Ltd.

Williams, V. (1984) Employment implications of new technology, *Employment Gazette,* May, pp. 210–215.

Winograd, T. (1972) *Understanding Natural Language,* Academic Press.

Winston, P. H. (1975) Learning structural descriptions from examples, in *The Psychology of Computer Vision,* McGraw-Hill.

Winston, P. H. (1977) *Artificial Intelligence,* Addison-Wesley.

Wood, B. O. and Fugelso, M. A. (1983) MCL, the manufacturing control language, *Proc. 13th Internat. Symp. Industrial Robots.*

Yakimovsky, Y. and Cunningham, R. (1978) A system for extracting three dimensional measurements for a stereo pair of TV cameras, *Computer Graphics and Image Processing.*

Yonemoto, K. (1981) The socio-economic impacts of industrial robots in Japan, *The Industrial Robot,* December, pp. 238–241.

Index